Resolving Environmental Conflicts

Social–Environmental Sustainability Series

Series Editor

Chris Maser

For more information about this series, please visit https://www.crcpress.com/Social-Environmental-Sustainability/book-series/CRCSOCENVSUS

Resolving Environmental Conflicts

Principles and Concepts

Third Edition

Chris Maser and Lynette de Silva

CRC Press
Taylor & Francis Group
Boca Raton London New York

CRC Press is an imprint of the
Taylor & Francis Group, an **informa** business

CRC Press
Taylor & Francis Group
6000 Broken Sound Parkway NW, Suite 300
Boca Raton, FL 33487-2742

First issued in paperback 2022

© 2019 by Taylor & Francis Group, LLC
CRC Press is an imprint of Taylor & Francis Group, an Informa business

No claim to original US Government works

ISBN 13: 978-1-03-247556-1 (pbk)
ISBN 13: 978-1-138-49882-2 (hbk)
ISBN 13: 978-0-429-20014-4 (ebk)

DOI: 10.1201/9780429200144

Library of Congress Cataloging-in-Publication Data

Names: Maser, Chris, author. | de Silva, Lynette, author.
Title: Resolving environmental conflicts / Chris Maser, Lynette de Silva.
Description: Third edition. | Boca Raton : Taylor & Francis, a CRC title, part
of the Taylor & Francis imprint, a member of the Taylor & Francis Group,
the academic division of T&F Informa, plc, 2019. | Series: Social environmental
sustainability | Includes bibliographical references and index.
Identifiers: LCCN 2018057159| ISBN 9781138498822 (hardback : alk. paper) |
ISBN 9780429200144 (ebook : alk. paper)
Subjects: LCSH: Environmental policy. | Environmental protection. |
Environmental mediation.
Classification: LCC GE170 .M369 2019 | DDC 333.7—dc23
LC record available at https://lccn.loc.gov/2018057159

Visit the Taylor & Francis Web site at
http://www.taylorandfrancis.com

and the CRC Press Web site at
http://www.crcpress.com

Dedication

We respectfully dedicate this book to students

of conflict resolution worldwide. We commend

you for your courage to accept the responsibility

of helping transcend our universal, conflict-

prone social conditioning and thereby honor

cooperation among people of diverse cultural

backgrounds. We recognize and applaud

your contribution to social-environmental

sustainability for the benefit of all generations.

Contents

Section II   The Legacy of Resolving Environmental Conflicts

Foreword

There are a number of schools of thought in the mediation/facilitation world about, well, just about everything:

- Should mediators be trained in the topic they are mediating or not? Some argue no, what one needs to know is how to bridge human divides, and the topic itself is not central. Some argue yes, especially when the subject matter is especially complex, and that without topical expertise, one will likely miss opportunities for agreement.

- Is a successful mediation about the results or the process? Those who argue the former often suggest that whatever feathers are ruffled will smooth out with time, especially if the results are desirable to all. Those who argue the latter suggest that without healing relationships through the process, any agreement may be too fragile to last.

On these and a host of other complex dichotomies, Chris Maser and Lynette de Silva seem to answer simply yes—all sides have value, and the dichotomies are mostly false. Perhaps because of their vast experience, both starting as scientists, then evolving to roles as mediators/teachers/trainers/researchers, their contribution with this work is not to advocate for any particular school of thought, but rather for the value in all sides of each complex setting, to be appreciated rather like the facets of a diamond.

Maser and de Silva weave together a number of important strands—environmental sustainability, resource conflict management, and the art of teaching—any of which could be a book by itself. Their contribution is not only in each of the topics—each of which is handled thoughtfully and with nuance—but also more so in the linkages between each strand, which seem to be where they find the sizzle. Just read these sentences, so emblematic of their approach:

> The aim of mediation, as we practice it, is to help parties become better human beings by stimulating growth in personal consciousness, thereby transforming human character, which results in parties finding genuine solutions to their real problems. In addition, the private, nonjudgmental, noncoercive character of such mediation can provide disputants a safe haven in which to humanize themselves, despite the disputants having started out as fierce adversaries. This safety helps people feel and express varying degrees of understanding and concern for one another as they grow toward greater mutual understanding and compassion, despite their disagreement.

The reader finds this exquisite blend of the physical, emotional, mental, and spiritual worlds throughout. This wonderful work is part description of some of the world's most seemingly intractable problems, part ethical polemic, part how-to manual, and, importantly, part how-to-teach manual. The authors are not satisfied with simply describing what is wrong with our relationships with each other and with our environment, and they do not stop when you personally are committed to action, but they will also give you both the motivation and the tools to help others learn how to do their parts in healing these relationships as well.

This is an important, readable, thoughtful book. As they suggest, "Ways will be explored through teaching and practice that will enable society to replace nonconstructive patterns and incorporate more meaningful connections"—aspirations that are sorely needed for our difficult times, and these are the guides to help us on our way.

Aaron T. Wolf
Professor of Geography, Oregon State University
Corvallis, Oregon

Series Preface

There are two primary emotions: love and fear. All other emotional expressions are merely aspects of these two. Kindness, compassion, and patience are the hallmark of love, while impatience, anger, and violence are the stamp of fear. Thus, where unconditional love dwells, there dwells also peace, contentment, and harmony—both inner and outer—wherein fear cannot endure. Where fear dwells, there is discord, discontent, and conflict, wherein peace cannot be found.

Conflict—like cooperation—is a choice, but one born out of one's fear of being out of control, whether of one's physical life, financial security, personal identity, or coveting someone else's possessions, such as land and its resources. Moreover, the dynamics of conflict are essentially the same, whether interpersonal, intertribal, international, or interreligious. Strife, after all, is dependent on the notion of inequality: I'm right; you're wrong. I'm superior; you're lesser. I belong; you don't. This is mine to do with as I wish; it's not yours—hands off. I want what you have, so give it to me or I'll take it.

The challenges we humans face in today's world are the result of unconscious, competitive, conflict-prone social conditioning, which begins at birth and ends at death. There is, however, no such thing as "right" or "wrong" in the universe, because the universe is an all-encompassing, holistic relationship based on eternal creation and novelty: All change is impersonal, neutral, and irreversible, despite the outcome.

Social conditioning, on the other hand, creates myriad perspectives that, in turn, spawn infinite, personal perceptions (human values), each accepted as "the truth"—from a certain point of view. The paradox is that everyone is right from his or her vantage point, which creates a venue of "right, right, and different." So, the question (and the heart of conflict resolution) becomes, How do we negotiate the differences while honoring one another's perceptions?

This being the case, resolving a conflict is based on the art of helping people with disparate points of view find enough common ground to ease their fears, sheath their weapons, and listen to one another for their common good, which ultimately translates into social-environmental sustainability for all generations. As it turns out, people agree on virtually 80 percent of everything—unbeknownst to them—and disagree on 20 percent, which becomes the sole focus of their dispute. If, therefore, disputants can be helped to see and move toward the predominance of their agreement, the differences ensconced in their quarrel are more easily negotiated. Ultimately, however, it is necessary for the participants to formulate a shared vision toward which to strive, one that accommodates the personalized perceptions to everyone's long-term benefit. Only then can the barriers among disputants dissolve into mutual respect, acceptance, and potential friendship—only then is a conflict truly resolved.

Chris Maser, Series Editor

About the Authors

Chris Maser spent over 25 years as a research scientist in natural history and ecology in forest, shrub steppe, sub-arctic, desert, coastal, and agricultural settings. Trained primarily as a vertebrate zoologist, he was a research mammalogist in Nubia, Egypt (1963–1964) with the Yale University Peabody Museum Prehistoric Expedition and a research mammalogist in Nepal (1966–1967), where he participated in a study of tick-borne diseases for the US Naval Medical Research Unit 3 based in Cairo, Egypt.

He conducted a 3-year (1970–1973) ecological survey of the Oregon coast for the University of Puget Sound, Tacoma, Washington. He was a research ecologist with the US Department of the Interior, Bureau of Land Management, for 13 years, the first 7 (1974–1981) studying the biophysical relationships in rangelands in southeastern Oregon and the last 6 (1982–1987) studying old-growth forests in western Oregon. He also spent a year as a landscape ecologist with the US Environmental Protection Agency (1990–1991).

Today, he is an independent author as well as an international lecturer and a facilitator in resolving environmental conflicts, vision statements, and sustainable community development. He is also an international consultant in forest ecology and sustainable forestry practices.

He has written over 290 publications, including 43 books that he has written, coauthored, or edited. His books are in libraries in 78 countries, including the United States and Canada.

He has lived, worked, consulted, or lectured in Austria, Canada, Chile, Egypt, France, Germany, Japan, Malaysia, Mexico, Nepal, Slovakia, Switzerland, and various settings in the United States.

Lynette de Silva directs the Program in Water Conflict Management and Transformation at Oregon State University. She teaches courses in water conflict management and water resources management and has acted as a consultant to the United Nations Educational, Scientific, and Cultural Organization (UNESCO), offering training to senior water professionals. Over the past 20 years, she has worked in areas emphasizing water resources and land management practices.

Acknowledgments

I (Chris) am once again extremely grateful to my wife, Zane, for her patience with me during the many hours working on this book that took precedence in our daily life.

I am also grateful for the honor of having Lynette de Silva join me in this endeavor, adding her skills, as a teacher of conflict resolution in water management, to smooth the path for humanity's peaceful use of Nature's irreplaceable water worldwide. Working with her is an inspiration.

To Aaron T. Wolf, I extend my sincere thanks not only for his clear, concise, and helpful review of the manuscript but also for his willingness to write a superb Foreword. To Janine Salwasser, I extend my sincere gratitude for her clear, innovative, and helpful review, which clearly improved our book.

Chris Maser, I (Lynette) thank you for the grand invitation to write with you! This gesture is a true compliment and is absolutely welcomed. It has been a pleasure to work with you. The journey was sometimes daunting, but always joyous. I do hope you know that I relish our conversations immensely.

Janine Salwasser of the Institute for Natural Resources, I so appreciate you devoting some of your summer to reviewing this book. Your constructive feedback and invaluable insights regarding the content and rearrangement of chapters has made for a better read. And, the addition of your situation map adds great value. Thank you.

Aaron T. Wolf, it has been a delight, professionally and personally, to work with you over the past 12 years. I am profoundly appreciative of the opportunities you have afforded me to co-direct Oregon State University's Program in Water Conflict Management and Transformation, co-teach, and collaborate on projects with you. My sincere thanks for your generosity and unwavering support. Having you review the manuscript and provide such a stellar Foreword means a lot to me. Thank you.

From Oregon State University, we recognize Nancy Steinberg for the early read and constructive suggestions regarding my Introduction and acknowledge David Reinert for assistance with redrafting the situation map.

I am immensely grateful to Zane Maser for so graciously agreeing to proofread this book. I do hope you know that I value your close and careful reading, and attention to details. And, I simply love the photograph you shot, that is utilized on the book cover.

Susan Eriksson and Janine Salwasser, you both bring added perspective and insight. So, I am doubly appreciative of your praise and kind endorsement.

To my parents, Gwendoline and Cyril Oscar Lucas, and my brother, Kwesi, I feel so fortunate to have come from such a nurturing and loving home. To my husband and mate, Shanaka de Silva, and my daughter, Chiara, I cherish your love and support.

Finally, we offer special thanks to Irma Britton, who asked for the third edition of this book; Kate Brown, one of the best copyeditors; and Marsha Hecht for her wonderful attention to details, as production manager. This is book, as always, is a team effort requiring everyone's cooperation, which is wonderfully demonstrated by our CRC team. Thank you.

Introduction

We are now in the twenty-first century, a century in which once-abundant natural resources are rapidly dwindling while the human population of the world continues to grow at an exponential rate. We, the adult, decision-making citizens of this planet, must now address an ethical question for the present and all of the future: Do we humans living today owe anything to the current and future generations of life on Earth, both human and otherwise?

With this question in mind, it is important to consider the consequences of our social conditioning as they pertain to the ideology of our individual and collective actions, lest the impact of our present course continues unabated. Although nature is neutral in all outcomes of the eternal, irreversible process of change, we can—and progressively are—destroying the ability of countless biophysical systems worldwide to serve us with the products and services we require for a good and sustainable quality of life.

Transforming this disharmony between people and the planet requires understanding the forks in the metaphorical road of mediation: the left-hand fork of problem-solving (symptomatic approach) and the right-hand fork of growth in personal consciousness (systems approach). The path we choose to actually resolve an environmental conflict will require a renewed sense of personal commitment, one that causes us, as individuals and as a society, to act now for the social-environmental benefit of all generations. What direction, you might wonder, must our renewed sense of personal commitment take? Personal growth in conscious awareness is vital to the answer because, as Albert Einstein said, "We cannot solve our problems with the same thinking we used when we created them."[1] However, to actually elevate our conscious awareness, we must become students of processes and let go our advocacy of positions and embattlements over winning agreement with narrow points of view.

Our personal and social reticence to openly and honestly address this question calls to mind a salient paragraph from a speech Winston Churchill delivered to the British Parliament on May 2, 1935, as he saw with clear foreboding the onrushing threat of Nazi Germany to Europe and the British people:

> When the situation was manageable it was neglected, and now that it is thoroughly out of hand we apply too late the remedies which then might have effected a cure. There is nothing new in the story. . . . It falls into that long, dismal catalogue of the fruitlessness of experience and the confirmed unteachability of mankind. Want of foresight, unwillingness to act when action would be simple and effective, lack of clear thinking, confusion of counsel until the emergency comes, until self-preservation strikes its jarring gong—these are the features which constitute the endless repetition of history.[2]

Consider that civilizations have evolved by similar steps: the growth of technology through discoveries and inventions, an increasing sense of environmental control, and social advancement through the ideas of government, family, and property, with all based on a slow accumulation of experiential knowledge. Namely, the arts of subsistence and the achievements of technology can be used to distinguish the periods of human progress.

People lived by gathering fruits and nuts; learned to hunt, fish, and use fire; invented the spear and atlatl and then the bow and arrow. They developed the art of making pottery, learned to domesticate animals and cultivate plants, began using adobe and stone in building houses, and learned to smelt iron and use it in tools. At some point, the irrigation ditch was devised, fostering the spread of agriculture, which not only carried human communities through difficult times but also enabled them to compete with one another by inhabiting new areas. Finally, what we call civilization began with the invention of a written language, culminating in all the wonders and many social-technological tragedies of the modern era.

There is a natural progression in civilizations of birth, maturation, and demise. The last is brought about by uncontrolled population growth, exemplified by current data, that outstrips the source of available energy, be it loss of topsoil, deforestation, or the continued despoliation of freshwater and the oceans. In olden times, however, survivors could move to less populated, more fertile, unpolluted areas as their civilizations collapsed. Today, there is nowhere left on Earth to go. Moreover, modern technology is both facilitating the overexploitation of Nature and polluting Planet Earth in the process, as well as dramatically altering the global climate for centuries to come.

Yet, having learned little or nothing from history, as Churchill pointed out, our society is currently destroying the very environment from which it sprang and on which it relies for continuance, thereby limiting our worldview and our evolutionary human existence to the biophysical, emotional, and intellectual realms—impeding our full potential. With this in mind, one might wonder what lies beyond our current notion of society and humankind? Are there other realms and other ways of being that humans could be considering and incorporating? Is the exploration of outer space part of it, as is so often stated? On the other hand, might it include inner space, the conquest of oneself? As the Buddha said, "Though he should conquer a thousand men in the battlefield a thousand times, yet he, indeed, who would conquer himself is the noblest victor."[3] In short, is our worldview too narrow?

In the material world, self-conquest means overcoming our materialistic social conditioning to bring our thoughts and consequential behavior in line with Nature's immutable biophysical laws governing the world in which we live. In the spiritual realm, self-conquest means disciplining our thoughts and behaviors in accord with social truths handed down throughout the ages, such as *treat others as you want them to treat you* and *you will reap whatever you sow.*

The outcome of self-conquest would represent an essential step toward social-environmental sustainability. This is the frontier beyond self-centeredness and greed-induced conflict, which destroys human dignity, degrades the productive capacity of our global ecosystem, and forecloses options for all life for all time.

To fulfill our obligation as social-environmental trustees for both the children we bring into the world and for all life requires fundamental changes in our personal social/religious conditioning, the cause of virtually every social-environmental conflict. The required changes demand choices different from those we have previously made, which means thinking and acting anew. But, "a great many people," as American psychologist William James observed, "think they are thinking when they are merely rearranging their prejudices."[4]

For example, under the administration of President Donald Trump the Environmental Protection Agency plans to roll back the Clean Power Plan, a major Obama era climate rule: "Perhaps no single regulation threatens our miners, energy companies and workers more than this crushing attack on American industry," Trump said in March 2017, referring to President Barack Obama's plan to slow the effects of global warming for the benefit of all generations worldwide.[5]

To change anything, we must, through the choices we make, reach beyond where we are, beyond where we feel safe. We must dare to move ahead, even if we do not fully understand where we are going or the price of getting there, because perfect knowledge will always elude us. Furthermore, as previously stated, we must become students of processes and let go our advocacy of positions and embattlements over winning agreement with narrow points of view. This is important because our ever-increasing knowledge rapidly outstrips the ability of our current paradigm, based on old knowledge, to explain the new in terms of the old.

True progress toward an ecologically sustainable environment and an ethically just society will be initially expensive in both money and effort, but in the end, it will not only be mandated by shifting public values but also be progressively less expensive over time. The longer we wait, however, the more deleterious will become the environmental conditions, such as the growing effects of global climate change on both regional temperatures and the dwindling availability of fresh water for human use. The growing uncertainty of locally available fresh water is already causing an increasing number of worldwide disputes based on competition for this inviolable requirement of life.

No biological shortcuts, technological quick fixes, or political hype can mend what is broken. Dramatic, fundamental change is necessary if we are really concerned with enhancing and maintaining the quality of life in a sustainable manner. It is not a question of whether we can or cannot change, but one of will we change or won't we. Change is a choice, a choice of individuals reflected in the collective of society and mirrored in the social-environmental landscape throughout the generations.

Can social-environmental conflicts be resolved? Emphatically, yes. But, they must be grounded on the personal growth of the disputants to transcend the limitations of their present level of consciousness.

Environmental conflicts are created by the social choices people make and thus can be resolved by electing different choices with resolution so firmly in mind that it naturally leads to a shared vision of the future toward which to build. Because people are often consciously blind to the motives of their choices, some kind of process is needed to help resolve conflicts by overcoming blind spots, the first step toward a shared vision based on, and leading to, cooperation.

In this regard, we choose the term *mediation*, which comes from the Latin *medius* ("middle"), as opposed to the word *facilitate*, which comes from the Latin *facilis*, meaning "easy." Mediation, in the sense we use in this book, means to conduct a process of communication whereby people are assisted in freeing themselves from difficulties and obstacles in making decisions that either avoid or eliminate conflict by forging commonly held values into a shared vision toward which to build collectively.

Therefore, to resolve social-environmental conflicts, the mediation process not only must have the greatest and longest lasting personal and social effect possible but also be as healing as possible because outcomes of such conflicts are, above all, based on intergenerational social/religious conditioning. By this, we mean it is the present generation's responsibility to serve the future by protecting the sustainability of its array of social-environmental choices—not the future generations' responsibility to endow the present with permission to continue overexploiting already diminishing natural resources. Today, however, the children of all generations are the ignored, silent parties in myriad conflicts.

This book covers some of the basic concepts that, over the years, we have found vital in helping resolve conflicts by bringing them to closure in a shared vision of the future and then by implementing that vision as sustainably as possible. Although some important procedural aspects of mediation are discussed, there is a standard, well-documented literature of the generalized procedures that does not need to be repeated here. The discussion is therefore confined specifically to points concerning mediation as it applies to personal growth and thus resolution at the true causal level. But, if we were to write a book on the general procedures of mediation, it would sound like a combination of *As a Man Thinketh* by James Allen,[6] *The Tao of Leadership* by John Heider,[7] and *Leadership Is an Art* by Max DePree,[8] with a liberal dose of common sense, compassion, and humility.

For whom is this book written? It is written for anyone interested in creating or furthering social-environmental sustainability and furthering the course of peace at all levels of society. These people include, but are not restricted to, community leaders, such as city mayors, city councilors, county commissioners, and community volunteers; bioregional visionaries; indigenous Americans and Canadians; members of Shinto, the original religion of

Japan that worships the forces of nature; government agencies; conservation groups in the United States and abroad; and teachers and students at various levels of education. It is also written for professional mediators interested in helping people grow toward a higher level of consciousness and thus personal responsibility, as well as for those who are interested in becoming mediators. And, it is written for whoever is considering resolving a dispute, teaching dispute resolution, or leading a community in formulating a vision for its future.

Why is this book written? It is written to give people the necessary philosophical underpinnings for practicing the type of conflict resolution that not only settles a dispute but also heals the people. Although we always endeavor to leave behind a working knowledge of the process itself each time we mediate the resolution of a dispute, we cannot impart, in so short a time, that which has taken us much of our life to learn. It is therefore our hope that this book will encourage you—and would-be mediators springing up around the world—to consider a true healing approach to conflict resolution as an optimal way to achieve long-term social-environmental sustainability.

To become a mediator, one must learn to be a "sifter," taking something from here and something from there, which is incorporated into one's own style and culture. Sifting is essential to one's continual growth as a servant of the parties in any social-environmental conflict.

This book opens with a brief comparison of approaches to conflict resolution and a quick look at how we generally mediate disputes, which creates a context for what follows. Thereafter is an examination of what goes into the mediation process, beginning with the "givens" of any social-environmental conflict. The givens are those inviolable biophysical and social principles that must be understood, accepted, and acted on if a conflict is to be resolved. One of our main purposes is to help people understand the social-environmental consequences they place on all generations through decisions they make—beginning with their children and their children's children.

It is imperative that people think through the long-term outcomes of their decisions. We say this because children are one of the two silent parties in all social-environmental conflicts; the global ecosystem and its productive capacity are the other. All disputants must understand the social, environmental, and economic circumstances to which they are committing the future because if the outcome of a conflict is a deficit in terms of the children's future options, the productive capacity of the ecosystem, or both, it is analogous to "taxation without representation," and that goes against everything a true democracy stands for.

In our experience, an environmental conflict must be mediated toward its natural conclusion, a shared vision of a sustainable future toward which to build. Such a vision is the necessary outcome of every conflict resolution if society, as we know it, is to survive beyond the twenty-first century. This is a critical idea because parts are often mistaken for wholes, and ideas are often viewed as complete when in fact they are not. Such is frequently the case with the resolution of a conflict, when the goal is seen only as the solution of an immediate problem, which equates to "symptomatic" thinking.

A *symptomatic thinker*, by analogy, is one who goes to their doctor because they feel ill. In turn, their doctor tells them to get more exercise and lose 20 pounds. To which they respond, "Can't you just prescribe a pill so I can manage my condition? I don't want to change my lifestyle." If, on the other hand, we modern humans are to live with any measure of dignity, well-being, and social-environmental sustainability, the symptomatic rationale embedded in the social conditioning of our contemporary society must give way to a systems approach that recognizes and accepts the worldwide recip-rocal interactions among people with one another and between people and Nature's biophysical environment.

In contrast, a *systems thinker* sees the whole in each piece and is therefore concerned about tinkering willy-nilly with the pieces, while entrained in a management mentality, because they know such tinkering might inadver-tently upset the sustainable, productive function of the system as a whole. A systems thinker is also likely to see him- or herself as an inseparable part of the systemic whole, and thus a caretaker, instead of an independent, superior entity with a management mentality. A systems thinker is willing to focus on transcending the issue in whatever way is necessary to frame a vision for the social-environmental sustainability for all generations into the unseeable future.

With the preceding two paragraphs in mind, we use the growing water crisis as a unifying example to illustrate how a resolved conflict with diver-gent perspectives, and the myriad perceptions they generate, becomes an unconditional gift to all generations, indeed, all life.

To this end, it may come as a surprise to learn the quest for harmony among individuals and parties, and between humans and the environment, can be accomplished through effective use of alternative dispute resolution techniques. However, there are some unique aspects of the problems tackled here that make environmental conflict management an even more effective tool when appropriately employed.

Unlike any other resource, our most precious natural resources (air, water, and soil) are irreplaceable; thus, addressing environmental discord requires a nuanced approach. Our continual attempts to "manage" Earth's biophysi-cal systems are progressively, and *negatively*, impacting life, including our very own existence. Consider that humans need a delicately balanced atmo-sphere of 78 percent nitrogen, 21 percent oxygen, and 1 percent argon, car-bon dioxide, water vapor, and other gases.[9] Without this unique atmospheric composition, we can only survive for a matter of minutes. Moreover, consider further that humans can survive less than a week without drinking water.[10]

Can you think of any service that does not require water? Water is needed for ecosystem services, agriculture, energy "production," and municipal and industrial purposes, in addition to providing both aesthetic and spiritual value[11] (Figure 0.1). Because there is no substitute, the ascension and demise of nations

FIGURE 0.1
Baptism in the Jordan River, Israel. (Photograph by Lynette de Silva.)

and empires have been attributed to the control, or loss, of available water.[12] As 80 percent of Earth's freshwater flow consists of rivers that cross boundaries of one or more nations, as well as various cultures within the nations, impacting 40 percent of the world's population,[13] having to share water can, does, and has caused heightened tensions. Stark examples include disputes between Egypt and Ethiopia along the Nile River, among stakeholders in the Klamath Basin that straddles the California-Oregon border in the United States, and between countless individual landowners worldwide who are vying for water.

Regardless of the circumstance or scenario, most water issues, whether at the international, regional, or individual level, are related to quantity, quality, timing, and access.[14] Fluctuating in space and time, water moves through the hydrologic cycle but is neither created nor destroyed. Moisture in the atmosphere falls as rain and snow to Earth's surface; there, it percolates through soil to become groundwater, flows as runoff to wetlands and rivers, or converts to vapor via evaporation. Eventually, however, it all returns to the oceans from whence it came, only to cycle again.[15] For all intents and purposes, Earth's water stock is maintained at about 332,500,000 cubic miles (mi^3) or 1,386,000,000 cubic kilometers (km^3).[16] The same water that courses through the hydrologic cycle today was consumed by members of ancient civilizations and even the dinosaurs. With a global population of humans, expected to reach 9.8 billion by 2050 and 11.2 billon by 2100 according to the

United Nations,[17] increased demands on our limited water supply could lead to more contentious environmental issues and conflicts.

Global-environmental issues go beyond the use of freshwater. Environmental degradation includes pollution of water, air, and soil; deforestation; loss of biodiversity; and the anthropogenic effects of climate change, ocean acidification, and depletion of the ozone layer. This list is not exhaustive by any means. To address these crises, it is vitally necessary to find effective, sustainable approaches.

When United Nations Secretary-General Ban Ki-moon spoke at Stanford University in 2013, he echoed the need for a different and expeditious approach:

> In the next twenty years, the world will need at least 50 per cent more food . . . 45 per cent more energy . . . and 30 per cent more water. At the current rate, we will soon need two planet Earths. But we have only one planet. There can be no Plan B because there is no planet B. Both science and economics tell us that we need to change course—and soon.[18]

In addition, renowned astrophysicist Neil deGrasse Tyson was recently asked, "Given how we have managed to damage the Earth, should we be looking for other places to live?" He responded, in part, "If you had the power of geoengineering to terraform Mars into Earth, then you have the power of geoengineering to turn Earth back into Earth."[19] This profound statement means that if we choose, we have the ability to transform our world, the ability to change course.

Yes, changing course can include "geoengineering," but to be clear, multiple approaches are needed to address the global environmental concerns. Regarding water, this includes educational programs focused on increasing our awareness of the value of water; more effective techniques for the conservation of water and its application to agricultural, industrial, and household uses; as well as the technological innovations to provide cost-effective techniques of desalination to make brackish and salt water usable.

This book offers a change in course, a paradigm shift. Provided are approaches that come from the field of environmental conflict management and transformation that are grounded in spiritual traditions. The approaches explore and address our whole being: physical, emotional, intellectual, and spiritual. We map pathways to collectively constructing a relationship-centric future, fostering healthier interactions with one another and more harmonious interactions with the planet. In essence, these approaches provide a foundation on which to build trust, skills, consensus, and capacity. If implemented effectively, this approach could be a "game changer."

As we write this book, we introduce our personal experiences by saying the following: I (Chris) found that . . . , or I (Lynette) teach this . . . . These examples are to help us recognize that the resolution of every conflict is a unique journey based on similar processes and novel outcomes.

Discussion Questions

1. What is the difference between symptomatic thinking and systemic thinking, and why is the distinction important in resolving an environmental conflict?

2. In resolving an environmental conflict, which is the most important, symptomatic thinking or systemic thinking?

3. How and why do the decisions made by adults affect all generations?

4. What do you think it means to resolve a conflict?

5. Is there a particular question you would like to ask?

Endnotes

1. BrainyQuote. Albert Einstein Quotes. https://www.brainyquote.com/authors/albert_einstein (accessed October 16, 2017).
2. Robert R. James, Ed. *Winston S. Churchill: His Complete Speeches, 1897–1963*. Chelsea House, New York, 1974, Vol. 6, p. 5,592.
3. *The Teaching of Buddha*. Bukkyo Dendo Kyokai, Tokyo, Japan, 1966.
4. Richard Alan Krieger. *Civilization's Quotations: Life's Ideal*. Algora, New York, 2002. 344 pp.
5. Stephanie Ebbs. How Trump's Decision to Roll Back the Clean Power Plan Could Affect the Environment. October 10, 2017. http://abcnews.go.com/Politics/trumps-decision-repeal-clean-power-plan-impact-environment/story?id=50389019&cid=clicksource_4380645_4_film_strip_icymi_hed (accessed October 10, 2017).
6. James Allen. *As a Man Thinketh*. Grosset & Dunlap, New York, 1981.
7. John Heider. *The Tao of Leadership: Leadership Strategies for A New Age*. Bantam Books, New York, 1986. 167 pp.
8. Max DePree. *Leadership Is an Art*. Dell Trade Paperback, New York, 1989. 148 pp.
9. Arnett, P. What Are the Three Most Abundant Gases in the Earth's Atmosphere? Updated April 25, 2017. http://sciencing.com/three-abundant-gases-earths-atmosphere-7148375.html (accessed September 20, 2017).
10. Donald B. Aulenbach. Water—Our Second Most Important Natural Resource. *Boston College Law Review*, 9(1968):535.
11. Aaron T. Wolf, Ed. *Hydropolitical Vulnerability and Resilience Along International Waters: Asia*. UN Environment Programme, Nairobi, 2009.
12. Rajendra Pradhan and R. Ruth Meinzen-Dick. Which Rights Are Right? Water Rights, Culture, and Underlying Values. In: Peter G. Brown and Jeremy J. Schmidt, Eds., *Water Ethics: Foundational Readings for Students and Professionals*. Island Press, Washington, DC, 2010, pp. 39–58.
13. Wolf, *Hydropolitical Vulnerability*.

14. Aaron T. Wolf, Annika Kramer, Alexander Carius, and Geoffrey D. Dabelko. Managing Water Conflict and Cooperation. In: Worldwatch Institute, *State of the World 2005: Redefining Global Security*. Worldwatch Institute, Washington, DC, 2005, pp. 80–208. http://www.transboundarywaters.orst.edu/publications/abst_docs/wolf_sow_2005.pdf (accessed September 20, 2017).

15. Chris Maser. *Interactions of Land, Ocean and Humans: A Global Perspective*. CRC Press, Boca Raton, FL, 2014. 308 pp.

16. Thomas V. Cech. *Principles of Water Resources: History, Development, Management, and Policy*. 3rd ed. Wiley, New York, 2010. 546 pp.

17. United Nations, Department of Economic and Social Affairs, Population Division. World Population Prospects: The 2017 Revision, Key Findings and Advance Tables. https://www.compassion.com/multimedia/world-population-prospects.pdf (accessed September 20, 2017).

18. Ban Ki-moon. Remarks at Stanford University on January 17, 2013. https://www.un.org/sg/en/content/sg/speeches/2013-01-17/remarks-stanford-university (accessed September 21, 2017).

19. Nicola Davis. Neil DeGrasse Tyson: "I Think the Things You Might Think Up in a Bar" [Interview]. *Guardian and Observer,* October 30, 2016. https://www.the-guardian.com/science/2016/oct/30/neil-dregrasse-tyson-astrophysics-mars-exploration (accessed September 20, 2017).

Section I

Mediating Environmental Conflicts

1

Approaches to Mediation

Mediation (facilitation, according to Bush and Folger) is generally understood as an informal process in which a neutral third party, one powerless to impose resolution, helps disputing parties seek a mutually acceptable settlement. As such, mediation has within itself the unique potential to raise the level of consciousness in disputants, which engenders personal growth by helping them, in the very midst of conflict, to wrestle with difficult inner and outer circumstances so they might better understand cause and effect, assume greater ownership of personal responsibility, and thereby bridge human differences.[1]

Nevertheless, the unique potential of mediation to achieve a higher level of consciousness is receiving less and less emphasis in practice. This potential is therefore seldom realized, and when it is, it is generally serendipitous, rather than the result of the mediator's purposeful efforts. Why? The answer lies in which of the two forks in the metaphorical road of mediation is chosen: the left-handed fork of problem-solving (symptomatic approach) or the right-handed fork of growth in personal consciousness (systems approach).

Forks in the Metaphorical Road of Mediation

The problem-solving approach to environmental conflicts is basically a business-oriented approach that emphasizes the capacity of mediation to find solutions that generate mutually acceptable settlements, almost always for the immediate benefit of adult humans, regardless of the effect of the settlement on children or the productive capacity of the environment. Mediators in this approach often endeavor to influence and direct disputants toward a settlement in general, and even toward the specific terms of a settlement, as I (Chris) have witnessed more than once with respect to a government-sponsored mediation concerning the use of public lands by the timber industry and the livestock industry.

As mediation has evolved, the problem-solving approach has been increasingly emphasized, to the point that this kind of directed, settlement-oriented outcome dominates the current movement. The premise of the problem-solving approach is to maximize the greatest possible immediate satisfaction of individuals engaged in a conflict. But, as author Gail Sheehy pointed out, "Human institutions prepare people for continuity, not for change." To us, therefore, the limitations inherent in the problem-solving approach are precisely its narrowness in scope; rigid focus on quick-fix, quantifiable outcomes; and the increasing attempt to eliminate risk, all symptoms of its growing institutionalization.

Designer Milton Glaser captured well our concern with the uncritical institutionalizing of professionalism in mediating the resolution of disputes when he said, "Professionalism really means eliminating risk. Once you become good at something, everyone wants you to repeat it over and over again. But the more you eliminate risk, the closer you come to eliminating the act of creative intervention."

In contrast to the problem-solving approach, raising the disputing parties' level of consciousness emphasizes the capacity of mediation for personal growth, which is embodied in the ability to accept risk. Mediators therefore concentrate on helping parties empower themselves to define the issues and decide the settlement on their own terms and in their own time through a better understanding of one another's perspectives.

To accomplish this increased awareness, mediators avoid the directedness associated with the problem-solving approach. Equally important, mediators help parties recognize and capitalize on the opportunities for personal growth inherently present in a conflict. This does not mean that satisfaction and fairness are unimportant; rather, it means that personal growth and more enlightened conduct are even more important.

The aim of mediation, as we practice it, is to help parties become better human beings by stimulating growth in personal consciousness, thereby transforming human character, which results in parties finding genuine solutions to their real problems. In addition, the private, nonjudgmental, noncoercive character of such mediation can provide disputants a safe haven in which to humanize themselves, despite the disputants having started out as fierce adversaries. This safety helps people feel and express varying degrees of understanding and concern for one another, as they grow toward greater mutual understanding and compassion, despite their disagreement.

The most important aspect of mediation is its ability to strengthen people's ethical resolve and their ability to handle adverse circumstances beyond the immediate conflict. It therefore transforms society for the better by bringing out the intrinsic good in people.

In mediation, it is not so much the procedural aspects that make the difference, but rather the nature and emphasis of the mediator's content within its context. As you read this chapter, remember that we are each a single person, and the way we mediate an environmental conflict or teach how to resolve such a dispute does not make it the only way or even the "right" way. What we offer is the best way we have learned from personal experience.

The Mediation Process, as Chris Practices It

I am deeply concerned with the philosophical foundation of the mediation I practice because, while working overseas, I have lived under both a ruthless dictator and a Communist regime that were, at best, indifferent to human life. From these experiences, it is clear that coercion of any kind settles no differences and lays to rest no issues. It only degrades human beings, trashes their dignity, and thereby steals hope from their souls.

I did not set out to become a facilitator. In the beginning, I merely noticed that environmental conflicts grew out of materialistic, human desires that were incompatible with the sustainable capacity of the ecosystem in question. This situation was compounded as questions concerning social values were increasingly subjected to "objective" scientific study; to derive objective scientific data; to provide objective scientific answers. Yet, despite millions of dollars and thousands of person-hours devoted to such study, intrinsic cultural values and objective scientific data remain miles apart. Why? This is the case because human objectivity is nonexistent due to the fact that every question is, by definition and intent, subjective.

The original idea, therefore, was to help bridge this chasm by presenting participants with the ecological concepts (based on the best available data) and the social concepts (based on their expressed cultural values) within the context of systems thinking. My sole intent was to help participants understand the information from an interactive, interconnected, interdependent, social-environmental-systems point of view, encompassing the past, present, and future. In this way, they could expand their common frame of reference in preparation for someone else to mediate the resolution of their dispute.

It was always emphasized that the data presented were the most up to date that I was aware of, but that *neither I nor anyone else knew what was "right" or had "the" answer*. Over time, and much to my surprise, participants began asking me to stay with them and guide the entire mediation process. Having no training and no idea what I was doing, I reluctantly agreed, but limited myself to working on environmental conflicts.

All I had in my favor was an undying belief (1) in the inherent goodness of people; (2) that people's blindness—their lack of conscious awareness—was born of ignorance, not malice; (3) that each person does the level best they know how to do at all times; (4) that given a place in which to be safe, an empathetic ear with which to be heard, and the empowerment with which to overcome fear and ignorance, people could and would change their behavior for the better; (5) that there was no turning back once a person started down the path of personal growth, with its increasing sense of freedom through self-control and self-direction; and (6) that society as a whole is lifted up each time an individual consciously grows and matures emotionally, spiritually, and intellectually.

My approach to mediation was intuitively transformative because it is vitally important to correctly identify the root cause to *eliminate* the symptom, rather than merely alleviate the symptom without touching its origin (the problem-solving approach). The approach I intuitively chose (1) assumes that human relationships take precedence over procedural outcomes; (2) opens people to greater compassion for one another; (3) allows people to argue for and protect one another's dignity; (4) is a meticulous practice of the best principles democracy has to offer, including patience; (5) balances intellect with intuition; (6) improves society by allowing people to grow emotionally, spiritually, and intellectually; (7) focuses on the cause of a conflict; (8) helps participants understand the consequences of their choices within the context of Nature's impartial, biophysical laws and their social-environmental reciprocity; (9) allows the outcome of a conflict to be decided solely by the participants, despite the fact that resolution may not take place until some months after mediation is completed; and (10) inspires the possibility of social-environmental sustainability.

While some mediators are fully versed on all aspects of a case before mediating, I intentionally go into mediation with as little specific knowledge of the conflict or the participants as possible. Although someone obviously contacts me about mediating the resolution of an environmental conflict, our conversation is kept to a minimum. Beyond that, I deal with as few participants as possible prior to the mediation process; if interaction is necessary, I avoid discussing the conflict. In this way, I am, as much as possible, unbiased during the process and detached from the outcome. This method has worked well over the years with such topics as the following:

1. Maintenance of a community's quality of life versus continual growth and commercialization by absentee companies;
2. Sustainable livestock grazing versus overgrazing on public lands;
3. Clear-cut logging in general versus soil stability and water quality;
4. Logging that excessively compacted the forest floor to the detriment of water infiltration and forest regeneration;
5. Logging versus protecting old-growth forests;
6. Logging that either did or would severely degrade a community's water supply;
7. Rights to the amount of water used by an upstream community versus a downstream community along the same small river;
8. Altering stream dynamics to the detriment of a community's immediate landscape; and
9. Emulating Nature's disturbance regime to maintain a particular forest composition versus public relations and a potential forest fire.

A sample of the groups involved in these conflicts includes the following: nongovernmental organizations (such as Malayan Nature Society, Penang, Malaysia; and Showa-Seitoku Memorial Foundation, Japan); government entities (like the federal Forest Service, Slovakia; city of Lakeview, south-central Oregon; US Bureau of Land Management in Idaho, Oregon, and Washington; and Mount Rainier National Park, Washington); and indigenous tribes (such as relating to the Bureau of Indian Affairs, Oregon; and Hoopa Valley tribe, California).

The Three-Part Mediation Process

Further, I only agree to mediate a conflict when the disputants have exhausted every other avenue of settlement and have come up empty handed. Only when they have reached "their wit's end" are they ready to listen and to change, which makes true mediation possible. There are three basic parts to the way I mediate the resolution of a dispute: (1) the introduction, (2) the body, and (3) the conclusion.

Introduction

The introduction serves several purposes, which collectively set the stage for the mediation process. The first is to establish common ground among the participants and between the participants and the mediator. I usually use cards with the participants' names on them to arrange seating, so that participants sit next to someone not of their choosing and begin to mix. Next, we take some time to introduce ourselves and share a little about our respective backgrounds, often including interests and hobbies.

At other times, the participants may form into pairs with their neighbors, take a few moments to learn about each other, and then take turns introducing each other. In this way, preliminary communication is initiated, which simultaneously starts to bring each person "out of the closet" through mutual participation, which captures and holds their attention and has within it the seeds of trust.

The introduction serves to clarify why the participants are taking part in the process and what each hopes to gain from it. At this time, we discuss whether all necessary parties are in attendance; if not, why not; and what can be done to rectify the situation. Here, my task is to help the participants develop an inclusive attitude by helping them understand why all parties are necessary to the process and its outcome.

Next, I help them develop a receptive attitude toward the mediation process itself. This is done by helping them understand what it may hold that is beneficial to them personally, such as learning how to use the democratic process as a tool of self-governance, as well as community governance.

The introduction also allows the disputants to learn what they can expect from me, what I expect from them, and what they can expect from one another. This is done in part by how I present myself as a person and in part by the collective establishment of rules of conduct, such as waiting one's turn to speak, being kind and polite at all times, and accepting one another's ideas without judgment.

Finally, the participants must understand that they will get as much or as little out of the mediation process as they are committed to putting into it. After all, it is theirs—not mine.

Body

The body is the main part of the mediation. When possible, I use a 3-day process. The first day is spent discussing what an ecosystem is, how it functions, the reciprocal nature of how and why we treat a system as we do, and how and why it responds as it does. By using slide presentations, it is easier for the participants to begin shifting their thinking prior to their explaining to me what the dispute is about because they know that under this circumstance I am as unbiased as possible.

This format works well, as evidenced by the article in the US Forest Service *Southwestern Region News*:

> Chris Maser . . . spoke to a group of Forest Service employees, preservationists, conservationists, and industrialists in Albuquerque [NM] August 9 [1988]. . . . Plans called for 30 to attend. There were 74 at the session with standing room only. . . .
>
> [Maser], who neither "preached" nor "put down" views of old growth, stressed repeatedly the need for rethinking positions, prior education, and conclusions concerning forests and old growth. A theme that was woven constantly through a several-hour monologue was that of discussion, definition, and consensus. The meeting and discussion of forest issues by those involved . . . is a necessity, he insisted. . . .
>
> Maser, with patience, painted a picture of the cycles of the forest, and the vital part of old growth. With meticulous care, he led the group through a labyrinth of long-term activity in the soils, which can, he insisted, produce a truly sustainable forest.
>
> A question and answer session followed the formal presentation. . . . It was notable that there were few pointed or conflict-prone questions. Reaction . . . from most participants was one of studied consideration of what had been presented, a reaction that we are convinced Maser was aiming for.[2]

The second day, with as much of a "systems" view as possible, we put the current conflict into a social-environmental context in the field, preferably in the area of contention. Going into the field is critical because it helps to make the abstract concepts of the first day into concrete experiences of immediate relevance. Here, the discussion begins by focusing on the teaching/learning of the first day, namely, on what an ecosystem is, how it functions, and the reciprocal nature of how and why we treat a system as we do and how and why it responds as it does.

During this time, each person in turn expresses a personal perception of the dispute from their understanding of how the ecosystem in question functions. The purpose is for each person to educate me about the dispute from their understanding of the whole and their perceived relation to the whole, based on their personal perspective from their lifelong social conditioning.

As each explanation unfolds, the person recounting it clarifies their own understanding of their perceptions, and the other parties hear for the first time the whole of someone else's story from that person's point of view. During this storytelling, I learn what the dispute is about because I hear it from various sides and am thus able to find common ground, differences, negotiable areas, quagmires, and hidden potentials for resolution.

Because the ecological condition of the resource(s), which is the focal point of the conflict, has a historical perspective, it is necessary to help participants examine this perspective. From our examination of the concrete historical and current perspectives, we progress to a more abstract, futuristic perspective. This perspective allows scrutiny of possible outcomes resulting from various kinds of decisions as each might affect the productive capacity of the environment, present and future. It is vital that the people involved be able to move from the concrete to the more abstract, based on their concept of current knowledge, if they are to craft a shared vision of the future as a resolution of their immediate conflict.

Toward this end, it is imperative to accept people where they are in terms of their understanding, which normally means using simple examples—often analogies—of how two or three components of a system might function together and then gradually expanding the examples to show how a more complex system might function as a whole. This includes helping participants understand such concepts as change in terms of (1) a continual, creative process of eternal, irreversible novelty; (2) self-reinforcing feedback loops; (3) isolated pieces versus interconnected, interactive functions; (4) the dynamic disequilibrium of an ever-changing system; and so on.

By accepting people where they are in their understanding, it is possible to help them move from a known point of departure, with respect to their perceived knowledge, toward new ideas and concepts while retaining their dignity intact. This process is greatly enhanced if one can lead people from more widely accepted ideas to those less widely held.

When I feel that I have an adequate understanding of the issue(s) and the participants seem ready (usually by the third day), we discuss the concept of a vision, goals, and objectives.[3] Once the participants have an understanding of these concepts, they begin to work out their vision and goals (it is not yet time for objectives), crafting them carefully on flip charts. Doing it this way, the vision and goals can usually be drafted and agreed to during the third day, as happened with the San Lorenzo Water District, Boulder Creek, California:

> The conflict over the disposition of the Waterman Gap forest (owned by the San Lorenzo Valley Water District) had raged for years, polarizing the community of Boulder Creek and destroying friendships. Although the

Waterman Gap advisory board members were in bitter conflict, after spending two days (in October 1997) with Chris Maser, the group reached a consensus and adopted a vision statement. The transformation was incredible. Not only did the board resolve its conflict, long-standing wounds were healed, and the group has since been able to work together constructively with shared goals and visions.

Since that time, I have conducted extensive research about environmental conflict resolution. After meeting with other mediators and witnessing other mediation sessions, I recognize the uniqueness of Chris Maser's approach—and, comparatively, how effective he is. Chris has facilitated more than 50 environmental conflict resolutions and has been successful in all of them. The process he uses does not solve a problem, it changes the way participants perceive and relate to each other, so people who were once combatants can work together in the future. Residents in Whitethorn, California, for example, have asked Chris to return to help them develop a sustainable vision for their community.

The world needs this approach to its conflicts.

(For a more thorough discussion, see Katherine Knight. 1999. "Oregon's Chris Maser: No Formula, But Disputes Resolved in Days or Less." *Consensus*, 42:2, 7.)

—Katherine Knight, Santa Cruz, California.

Occasionally, however, this does not work. If the participants just do not agree, the disputing parties are separated and sent off by themselves, sometimes for a day or a couple of weeks. Each disputing party, which usually consists of a number of individuals, is tasked with crafting their own vision and goals. When they have completed the assignment, we reconvene, at which time each party shares its long- and short-term aspirations with the others, as happened with the Umatilla Forest Resource Council in Walla Walla, Washington:

In November 1987, the Umatilla Forest Resource Council (UFRC), a group of local citizens organized to address the planning process for the Umatilla Forest, invited Chris Maser to present a public address on sustainable forestry. At that time, he also agreed to conduct a half-day workshop for UFRC, helping us to focus our efforts on the process and on ways to approach forest planning.

When the Draft EIS [Environmental Impact Statement] for the Umatilla National Forest was released the following month, UFRC joined with Ralph Perkins, District Ranger on the Walla Walla Ranger District, and invited Chris Maser back. That time the invitation was to conduct a two-day workshop for a number of interested persons, all committed to involvement in the Umatilla Forest Plan. The group of thirty people was a cross section of Forest Service personnel, environmental activists, Confederated Tribes—Umatilla Indian Tribe, industry (private and corporate), Oregon and Washington Departments of Wildlife, Oregon Rivers Council, County government representatives, and other concerned citizens.

The primary objective of the workshop was to focus on interpersonal and philosophical foundations and the processes that provide the products on US Forest Service lands. Chris Maser worked from the premise that before we could hope to settle land management issues out of court, we must shift our focus from a product only orientation to a process orientation. For two days we worked, first in small groups and then together in a large group, learning to set a vision and goals and in the process learning to help each other define areas of agreement and areas that might need to be negotiated. The end product of the workshop was to begin framing a vision and goals for the forest, based on those areas of agreement. Chris wanted to create a workshop that would help us make a necessary shift in thinking and a change in the process in such a way that all parties involved might retain our dignity—the prerequisite for "winning."

The sessions were based on mutual learning, sharing, defining fear, expectations, hidden agendas, and values exercises. We dealt with change and paradigms and perhaps most importantly, with the concept of dignity in consensus groups. Strong points centered around listening actively—not listening is a form of violence and forming a rebuttle [sic] [before the other person is finished speaking] is a defense mechanism. We shared a common concern for decisions in land management and recognized that the outcome of our work would in the end become a human judgment decision.

With that beginning, the "Guiding the Course" group set ground rules, jelled the make-up of the group and began working together in an effort to develop a sound management plan for the Umatilla Forest. Four years and countless meetings later, a final plan was released. I can give no better endorsement for Chris Maser and his training sessions than to state that the group is still intact in 1994, with some changes in individual members, and is still meeting on a regular basis to continue monitoring "our" national forest. Our success has been measured not so much in land management decisions, although there have been great changes there as well, but in the continued respect and dignity we give to each other. We have learned to listen to each other, to respect each other's opinions, to be unafraid to offer a candid opinion and we are still committed to the process. Perhaps most important, we are friends.
  –*Shirley Muse, Walla Walla, Washington*

The purpose of one party presenting its vision and goals to the others is simply for the other people to help make sure—without judgment—that the stated vision and goals fit the agreed-on criteria. If they do not, the wording is corrected so that the criteria are in fact satisfied. Each party in turn presents its material, and each party in turn helps the others ensure that the criteria are met.

Once this process has been completed, all parties look for areas of overlap. I may help them with questions, a powerful tool when used wisely, because questions open the door of possibility. For example, it was not possible to go to the moon until someone asked the question: Is it possible to go to the moon?

At that moment, going to the moon became possible. To be effective, however, each question must (1) have a specific purpose; (2) contain a single idea; (3) be clear in meaning; (4) stimulate thought; (5) require a definite answer to bring closure to the human relationship induced by the question; and (6) relate to previous information.

For example, in a discussion about going to the moon, one might ask: Do you know what the moon is? The specific purpose is to find out if one knows what the moon is. Knowledge of the moon is the single idea contained in the question. The meaning of the question is clear: Do you or do you not know what the moon is? The question stimulates thought about what the moon is and may spark an idea of how one relates to it; if not, that can be addressed in a second question. The question, as asked, requires a definite answer, and the question relates to previous information.

Once the areas of agreement or willingness to compromise are found, they may constitute up to 80 percent or more of a common ground, and there may be little dispute left to negotiate. When this point has been reached, the parties are ready to conclude this phase of the mediation process.

Conclusion

In winding down this phase of the mediation process, the important elements of the dispute and its resolution are retraced, so the parties, having been consumed in the process, can now stand back and see in perspective how it works, which may give them a better understanding of the whole. This review both reinforces what they have learned and improves their retention of it for later reference. New ideas are not included at this time because they are likely to confuse the participants.

Finally, I must help them to determine what their next step is, usually another meeting to refine their initial draft of the vision and goals. They must decide how they want to do this and when. It is imperative, however, that they have their next meeting date set and committed to prior to adjourning.

Mediation in this way helps the parties create a shared vision and goals for a sustainable future in which all parties not only can benefit but also want to share. Only now do I consider the dispute largely, but not completely, resolved. Full resolution of an environmental conflict requires putting the shared vision into action as "one body."

> Diane Sawyer, the ABC news anchor covering the developing story [of the magnitude 9 earthquake that struck Japan on March 11, 2011] mentioned that, in the Japanese language, one of their characters means **"one body."** As she visited with numerous Japanese, they shared how they come together in solidarity and function as **a single body** in this sort of national crises, much like firefighters who hang on to their colleagues through the most perilous situations of life and death. *If one falls, they all fall. If you go down, I go down too.* I stand by you, and you stand by me. Thus, your home, life, and being are [as] precious to me

as mine are to you. Not a single incident of "'looting' [*was*] reported in Japan" [emphasis added].

"There is no question the Japanese respond well to this kind of catastrophe, but even if it looks remarkable from the outside, it's not new," said Carol Gluck, a professor of modern Japanese history at Columbia University's Weatherhead East Asian Institute. "It's not cultural or religious—it is a historically created social morality based on a response to the community and social order.". . .

"It's not that the Japanese are naturally passive and obedient," she said. "There is a historically created social value to it. People uphold it. It works." In other words, it is a form of social conditioning wherein members of the overall community look out for one another.[4]

With these paragraphs in mind, resolving an environmental conflict depends first on understanding the cause or causes of the conflict. Such understanding must uncover the chain of events set in motion by the participants' decisions, which in turn triggered cause-and-effect relationships within a range of alternative decisions and outcomes. My perception of a conflict must be as objective as possible and not based on judgment, as dictated by my social conditioning, and thus my standard of right or wrong.

Overcoming Animosity

Instead of developing an adversarial attitude, in my mediation style, I choose to create an environment in which all participants are respected and feel safe to express their opinions and concerns. It sometimes means being direct enough to voice my observations, even when they are not what the participants want to hear. But, I learned to do it in a way that is respectful and appropriate, not destructive and not in the form of a personal attack; this experience forms the basis of my mediation style, one of *respect for every participant.*

Engaging participants to work positively together can, at times, be a daunting task. In many instances, we bring a variety of barriers with us to the table, such things as company- and institutional-branded items (uniforms, caps, shirts). They are worn to meetings to clearly state, "This is who I am, and my position on this matter is clear." My first strategy as a mediator in these meetings is not to wear any identifying monogram. Instead of seeing me as "X from X agency," I want to be seen as a person, not a position. I also prefer to find neutral ground for the meeting, somewhere that is not the battleground itself.

One of my most valuable tools is the social gathering. When I want a group to work together and see each other as people, not positions, I plan a gathering that has a social context to it, such as a barbeque, open house with coffee and donuts, or a tour of a facility or field location with similar issues, followed by refreshments. By doing this, I have found that it not only sets the tone for the upcoming negotiations but also encourages participants to talk about themselves, as people, in a less formal, nonthreatening setting. I can tell you from personal experience that it is much harder to be angry with someone you have shared your children's baby photos with or joked with during a game of softball. Finally, such a gathering also gives me the opportunity simply to visit with the people, which allows them a chance of sensing who I am as a person. This is vitally important because people do not care how much you know until they first know how much you care about them, which is the essence of how Lynette teaches conflict management.

Conflict Management, as Lynette Teaches It

My goal in teaching is to provide skills that empower individuals to make meaningful linkages (through information, processes, or inspiration) that transform lives and global endeavors. I support efforts toward building a more inclusive world and sustainable future, and I support the idea that building the capacity for making wise decisions based on critical thinking is central to social-environmental sustainability. As such, I see the need for honing broad-based knowledge, collaboration, and lifelong learning.

Conflict management, as I teach it, is designed to hold a safe space for thinking and conversing on issues of water and environmental concerns. It is relationship-centric, with an emphasis on building constructive change toward healthier engagements and interactions among individuals of all racial, religious, and educational backgrounds, communities, and nations and a *preferred* interconnection with Mother Earth. The aim is toward achieving a transformational process, rather than simply resolving the situation at hand. We delve into underlying interactions and issues in a pluralistic approach, not one confined to a particular conflict. This training parallels mediation techniques and provides life-changing practices, so participants can learn through the experience of doing, with emphasis on hydro (water) considerations.

I teach conflict management through a course, Water Conflict Management and Transformation. Designed for everyday citizens, university students, environmental practitioners, and water diplomats, this course requires participants to have no prior knowledge of dispute resolution. It is currently offered through Oregon State University (OSU) online in the fall term; it is co-taught with Dr. Aaron Wolf (OSU geography professor, world-acclaimed water facilitator) as a 1-week workshop on campus each June. The workbook utilized during this training is *Sharing Water, Sharing Benefits: Working Towards Effective Transboundary Water Resources Management*.[5] This book is versatile in offering instruction for students as well as guidance and resources for instructors.

A Practical Course with Multidisciplinary Applications

My teaching is built on Wolf's "four stages of water conflict management framework,"[6] based on his study of over 140 treaties,[7] and founded on the works of Rothman[8-10] and Kaufman.[11,12] The four stages correspond to the four worlds' framework,[13] found in ancient spiritual traditions that focus on the physiological, emotional, intellectual, and spiritual fundamentals of the whole person (worlds), also identified through Maslow's hierarchy of needs.[14] Additionally, I draw from my own informal training (mediator instruction and application, meditation, conversational skills, and professional coaching); specialist environmental experiences; and personal reflections and encounters.

The Four Stages of the Water Conflict Management Framework

The four stages of water conflict management framework provide an approach within which to address conflicts from the adversarial, reflective, integrative, and action stages[15] (Table 1.1).

While the adversarial, reflexive, integrative, and action stages are typically enumerated for convenience in teaching, they are often concurrent in human behavior. Furthermore, when dealing with the differences of complex systems, these stages can apply equally within the perception of a single person (intrapersonal) or among the varied perspectives of people within a group (interpersonal).[16]

TABLE 1.1

Four Stages of Water Conflict Transformation

Negotiation Stage[a]	Common Water Claims[b]	Collaborative Skills[c]	Geographic Scope
Adversarial	Rights	Trust building	Nations
Reflexive	Needs	Skills building	Watersheds
Integrative	Benefits	Consensus building	"Benefit-sheds"
Action	Equity	Capacity building	Region

Source: This framework is built mainly on the works of Jay Rothman,[8-10]; Kaufman[11,12]; and Wolf.[6] This table is modified from Figure 6.1, in Jerome Delli Priscoli and Aaron T. Wolf. *Managing and Transforming Water Conflicts*. Cambridge University Press, New York, 2009, 354 pp.

[a] These stages build primarily on the work of Jay Rothman, who initially described his stages as adversarial, reflexive, and integrative (ARI).[8] When ARI become ARIA, adding action, Rothman's terminology[10] also evolved to antagonism, resonance, invention, and action. We retain the former terms, feeling they are more descriptive for our purposes.

[b] These claims stem from an assessment of 145 treaty deliberations described by Wolf.[6] Rothman[9] also used the terms *rights*, *interests*, and *needs*, in that order, arguing that "needs" are motivation for "interests," rather than the other way around as we use it here. For our purposes, our order feels more intuitive, especially for natural resources.

[c] These sets of skills draw from Kaufman,[11,12] who tied each set of dynamics specifically to Rothman's ARIA model in great detail, based on his extensive work conducting "Innovative Problem Solving Workshops" for "partners in conflict" around the world.

Adversarial Stage

The adversarial stage represents the most contentious exchange among a group of disputants. Demands are often made, but there is little to no civil discourse. Perceived historical events color the strained interactions that mire trust among disputants (Table 1.1). Claims are made for personal rights and ownership.[17] Prescriptive measures include honing the ability to listen,[18] building trust, and increasing awareness.[19]

Initially, the classroom seating is arranged in a semicircle for equal inclusivity of participants. Every effort is made to create a climate conducive to learning and forming healthy relationships. This includes a setting that is welcoming and nurturing for cordial rhetoric, where a richness of diverse perspectives (around water, environment, perceived needs, and interests) can be respectfully expressed; where fairness and equity are core classroom values; and where connecting with participants individually and collectively is the norm. Training then commences with students introducing themselves conventionally and through personal stories so they can experience the subtle differences an introduction can make.[20]

To support and stimulate healthy discussions among participants from different communities, cultures, and backgrounds, it is important to establish a common understanding for classroom engagement. So, through facilitation, participants agree on guidelines of acceptable behavior for the duration of training, such as placing electronic devices on mute, not interrupting someone speaking, and no smoking in the classroom. This becomes a class agreement that can be revisited and added to, if needed.

During this part of the training, presentations on water management and dispute resolution are conducted. Through exercises, participants experience the complexities of managing water and practice one-on-one interpersonal skills and effective communication, such as listening with patience. Contemplative practices (pauses of silence, meditation) are also employed at the beginning of each class and after breaks to bring greater clarity to the present moment.

Reflective Stage

In the reflective stage, there is a greater propensity toward listening and truly hearing motivations. In this way, actual context is given to biophysical requirements, rather than the personal claims surrounding past actions of opposing parties.[21] Here, emphasis is on continuing to build trust, along with furthering communication skills (Table 1.1). Tools toward constructive change in the classroom and in mediation include having parties recognize and validate the experiences of others and, when appropriate, engage in collaborative problem-solving toward mutual benefits.

One-on-one exercises continue to enhance listening. In these early activities, for example, discussions might be centered on a minor annoyance (a pet peeve), for which individuals take opposing positions. Each must give undivided attention to the other, taking turns listening to what is being said and paraphrasing the essence of what they heard. If further clarification is required, the process is repeated.

When the exercise is over, even though it was based on a fictitious scenario, it is good for the participants to *clear the air* by thanking those with whom they performed the exercise, just as sport competitors shake hands at the end of a game. This is relevant because, in the same way that active listening and healthy engagement can build trust in *real* negotiations, participants in these exercises are forming bonds of friendship.

These combined activities tend to expand our understanding of a given, contentious situation, shifting perspectives from an entrenched position to why a resource (greater allocation of water) is actually required. Regarding landscape space, there can be a broadening in outlook (property boundary to a neighborhood or a national border to an international watershed).[22] During this part of the training, relevant environmental case studies are shared, demonstrating real-world shifts from special interests to biophysical necessities.

Integrative Stage

When parties begin seeing themselves as members of something larger than themselves (a community, shared watershed, social-environmental system), they are entering the integrative stage.[23] This phase naturally allows for the broadening of perspectives, inclusivity, and resourcefulness. It requires cooperation with respect to water and approaches that benefit the stakeholders *at large*. This stage focuses on current and future requirements across an array of ecological, economic, and social fronts,[24,25] which might, for example, result in enhancing biodiversity, regional trade, and water security.[26]

With respect to consensus building (Table 1.1), participants in the classroom perform exercises that are group oriented (four to six people) and thus expand their individual perceptions. Restating a conflict in a new, but authentic, narrative can alter perceptions. And, field trips to engage participants with real stakeholders concerning environmental agreements/discourse in a community can be insightful. Moments spent silently communing with Nature can also prove invaluable. Both the field trips and outdoor contemplation take participants out of the classroom and can increase their "discernment and attentive capacity,"[27] allowing them to make new connections and achieve new insights (Figure 1.1). Other exercises (games that challenge assumptions and develop situation maps) can stimulate creative approaches to problem-solving. On the regional scale, in the classroom, participants learn the value of institutional capacity building and ways in which it can influence the resilience of a river basin.

FIGURE 1.1
Oregon State University students on a natural resources field trip at Whychus Creek, central Oregon, in the northwestern United States. (Photograph by Lynette de Silva.)

Action Stage

The operational component of the four stages focuses on action (Table 1.1), thereby ensuring that the equitable distribution of gains are both negotiated and realized *on the ground*.[28] To successfully execute these negotiations, new networks, policies, and institutions may be needed. Therefore, economic and legal mechanisms (funds and laws) are included for water, along with wider social and ecosystem concerns.[29] Such agreements, joint water projects, and collaborations inevitably strengthen social-environmental relationships and create a more resilient society.[30]

As the training draws to a close, just as within a real negotiation process, time is made for debriefing and ensuring that mechanisms are in place for a successful transition to the next steps, which may include processes for obtaining additional resources and support systems for oneself (participant) and others in a joint environmental quest.

Conclusion

This training in the four stages of water conflict transformation provides an introduction to environmental politics and the challenges of territorial boundaries present for allocation of water. Moreover, there are opportunities to examine interactions at the personal, regional, and international scales; to map social-environmental connectivity; and to study patterns of communication. We examine frameworks and ways to enhance our capacity through reframing issues. There are opportunities to apply these newly learned skills toward day-to-day communication through role-playing in fictitious scenarios. In this way, we explore paths to strengthen personal, regional, and institutional capacities.

The beauty of this training is that participants, in a safe and nurturing setting, are learning about processes by engaging in them. This multifaceted approach to learning also involves transformative and contemplative practices that can stimulate new relationships. Through the course design and hands-on nature of the material, the training stretches the fully engaged individual to communicate more effectively beyond cultural-social conditioning. It also empowers us to address new and changing situations that can resonate through our four worldviews (biophysical, emotional, intellectual, and spiritual). If practiced and honed in our everyday lives, these applications can effectively influence environmental conflicts. Throughout the chapters in this book, the essence of these stages of resolving social-environmental conflicts are explored in greater depth.

Discussion Questions

Discussion Questions from the Mediation Process, as Chris Practices It

1. While a variety of approaches to mediation exists, what might be gained by going into a mediation process knowing as little about the conflict as possible?
2. What is meant by "raising the level of consciousness"?
3. What does it mean in terms of conflict resolution?
4. What has raising people's level of consciousness got to do with a systemic vision as a way to achieve the culmination of a conflict?
5. Why is it important for the mediator to lay ground rules, discuss acceptable behavior, and ensure that agreed-to social boundaries are respected?
6. What is meant by "acceptable behavior"?
7. What is the difference between tolerance and acceptance? Which one is vital in resolving a conflict? Why is it critical?
8. Can you think of any relevant questions to discuss?

Discussion Questions from the Mediation Process, as Lynette Teaches It

1. Why is attention given to classroom seating? What relevance does this have?

2. Why is emphasis placed on building healthy relationships rather than solving the dispute?

3. How might personal introductions through stories be different from conventional forms of introduction?

4. Why are the participants in the classroom asked to establish their own class agreement?

5. Why do class activities begin with one-on-one participant interaction? Why not begin with larger group activities?

6. Why are building interpersonal skills and effective communication so crucial to conflict resolution?

7. With reference to collaborative skills, what is meant by "If practiced and honed in our everyday lives, these applications can prove impactful in environmental conflicts?"

8. What is meant by the idea that participants in the classroom are "learning about process, while engaging in them?"

Endnotes

1. Robert A. Baruch Bush and Joseph P. Folger. *The Promise of Mediation: Responding to Conflict through Empowerment and Recognition*. Jossey-Bass, San Francisco, 1994. 296 pp.

2. USDA Forest Service. Maser: A Man with a Mission. *Southwest Regional News*, 1998, pp. 6, 16.

3. Chris Maser. *Vision and Leadership in Sustainable Development*. Lewis, Boca Raton, FL, 1998. 235 pp.

4. The three paragraphs of the quotation are based on John Paul Lederach. *The Little Book of Conflict Transformation*. Good Books, Intercourse, PA, 2003. 74 pp.

5. Aaron T. Wolf (Ed.). *Sharing Water, Sharing Benefits: Working Towards Effective Transboundary Water Resources Management. A Graduate/Professional Skills-Building Workbook*. UNESCO, The World Bank, and Oregon State University, 2010.

6. Aaron T. Wolf. Criteria for Equitable Allocations: The Heart of International Water Conflict. *Natural Resources Forum*, 23(1999):3–30.

7. Ibid.

8. Jay Rothman. Supplementing Tradition: A Theoretical and Practical Typology for International Conflict Management. *Negotiation Journal*, 5(1989):265–277.

9. Jay Rothman. Pre-negotiation in Water Disputes: Where Culture Is Core. *Cultural Survival Quarterly*, 19(1995):19–22.

10. Jay Rothman. *Resolving Identity-Based Conflicts in Nations, Organizations, and Communities.* Jossey-Bass, San Francisco, CA, 1997. 224 pp. quotation for Note 4.
11. Edward Kaufman. Sharing the Experience of Citizens' Diplomacy with Partners in Conflict. In: John Davies and Edward Kaufman, Eds., *Second Track/Citizens' Diplomacy.* Rowman & Littlefield, Lanham, MD, 2002, pp. 183–222.
12. Edward Kaufman. Toward Innovative Solutions. In: Davies and Kaufman, *Second Track/Citizens' Diplomacy,* pp. 183–222.
13. Aaron T. Wolf. *The Spirit of Dialogue: Lessons from Faith Traditions in Transforming Conflict.* Island Press, Washington, DC, 2017. 205 pp.
14. Abraham Maslow. A Theory of Human Motivation. *Psychological Review,* 50(1943):370–396.
15. Wolf, Criteria for Equitable Allocations.
16. Ibid.
17. Jerome Delli Priscoli and Aaron T. Wolf. *Managing and Transforming Water Conflicts.* Cambridge University Press, New York, 2009. 354 pp.
18. Barbara Cosens, Lynette de Silva, and A.M. Sowards. Introduction to Parts I, II, and III. In: Barbara Cosens, Ed., *Transboundary River Governance in the Face of Uncertainty: The Columbia River Treaty.* A Project of the Universities Consortium on Columbia River Governance. Oregon State University Press, Corvallis, OR, 2012. 464 pp.
19. Delli Priscoli and Wolf, *Managing and Transforming.*
20. Michelle LeBaron. Bridging Troubled Waters: Conflict Resolution from the Heart. Jossey-Bass, San Francisco, 2002. 352 pp.
21. Delli Priscoli and Wolf, *Managing and Transforming.*
22. Ibid.
23. Ibid.
24. Claudia Sadoff and David Grey. Beyond the River: The Benefits of Cooperation on International Rivers. *Water Policy,* 4(2002):389–403.
25. Delli Priscoli and Wolf, *Managing and Transforming.*
26. Sadoff and Grey, Beyond the River.
27. Daniel P. Barbezat and Mirabai Bush. *Contemplative Practices in Higher Education: Powerful Methods to Transform Teaching and Learning.* Jossey-Bass, San Francisco, 2014. 231 pp.
28. Delli Priscoli and Wolf, *Managing and Transforming.*
29. Sadoff and Grey, Beyond the River.
30. Delli Priscoli and Wolf, *Managing and Transforming.*

2

Conflict and Cooperation Are Choices

Conflict is a choice of behavior. We resort to conflict in one way or another, at one level or another, because that is what we have been taught to do—our social conditioning, beginning as children. In this way, it becomes our inherent reaction that requires a systematic response. That is how we were taught to cope with change: those circumstances perceived as threatening to our survival. For example, we are waging a war against Nature, a war on cancer, a war on poverty, a war on obesity, a war on diseases of all kinds, a war on drugs, a war of the sexes, and even a war on violence. We are progressively making the world into an everyday combat zone with our thinking.

Looking around the world today, is it any wonder that various segments of the global society are blowing themselves to bits and in the process needlessly, recklessly squandering the natural resource base on which they and all future generations depend for survival? Children are being ushered into emotionally shattered lives, where their inner poverty will compound the outer poverty they face in the spiritual/cultural/economic chaos of disrupted lives; gutted cities; corrupt, power-hungry governments; and war-torn, fragmented landscapes. These generations may well grow up thinking that hatred and violent conflict are the norm, which continually fosters the unworkable paradigm of a black-and-white world in which

- I'm right and you're wrong.
- My religion is the only one that is right; thus, God is on my side, which makes you the infidel.
- You're either with me or against me.

Is such a future unavoidable? Must we increasingly become a world of victims in which there is no escape from an eternal dysfunctional cycle of abuse and combat? If abuse, along with the distrust and combat it engenders, is indeed the lot of humanity, then time and history will grind wearily on to the only social outcome possible, the ultimate destruction of human society, taking much of life on Earth with it. Is this the lesson human history will inevitably continue to teach, as each day's activities are recorded in the archives of eventide?

We think not. We see the world differently, and this transformation is predicated largely on three things: (1) We recognize, accept, and act on the notion that conflict is a choice, which means cooperation is also a choice, with differences resolved in peaceful ways; (2) we understand that the peaceful way lies in the art of mediation that fosters personal growth, with differences resolved through inner shifts in consciousness; so (3) we, as humans worldwide, develop the humility and wisdom to honor and abide first and foremost by Nature's biophysical principles: the rights of Nature.[1] Such shifts, which are core to the four stages of water conflict management (Table 1.1), alter the core perceptions from which conflict grows by increasing the social functionality of the participants and hence their tendencies toward peace and its social-environmental sustainability.

Simply put, mediation can help disputing parties mirror their fears and sense of vulnerability to one another, thereby coaxing from deep within a sense of compassion that transcends each person's individual fears. As fear is transcended, perceived differences among contestants' ideas usually become novel approaches to the commonalities of a shared vision. In fact, they often become a vision's source of strength.

There being no external fixes for internal fears, we must acknowledge that the cause of conflict is internal to the disputing parties themselves, as it always is to us. For example, children are taught by adults' behavior that interpersonal conflict is simply a condition of life, a necessity of survival. Watching my parents, I did, as a child, the only thing a child knows how to do: I (Chris) copied what I saw—the beginning of thoughtless social conditioning.[2]

Conflict is often assumed to be both open and visible because the term *conflict* brings to mind the images of war and abuse mentioned previously, as well as homegrown gang violence. As mediators, however, we must identify the hidden internal struggles, as well as the obvious external conflicts. We have experienced this assumption in resolving conflicts, when meetings seemingly go very well and agreement (or the appearance of it) brings the process to a pleasant conclusion. At the next meeting, we return to the issue and discuss it further, and at the next meeting, and the next, and so on. If you have been in this situation, you have seen the avoidance of actually confronting a conflict because of personal fear: If I speak up, will others dislike me? Be angry with me? Ignore me? Ridicule or threaten me?

However, there are "no enemies out there," only frightened people who feel the need to defend themselves from potential loss of what they think they must have to survive: control of their own lives as they perceive it. Thus, as mediators, it is imperative that we understand why a particular conflict is the chosen way of dealing with personal differences if we are to help the participants do anything more than temporarily alleviate the symptoms that festered into open dispute. To do so, we must understand something of the conflict's underlying nature, beginning with the notion of each person's "right" of survival, however that is defined.

What Is a "Right"?

In medieval literature, brave knights came from across the land to be considered for membership at the Round Table. King Arthur designed its circular shape to arrange the knights democratically and give each an equal position. When a knight was granted membership at the Round Table, he was guaranteed equal stature with everyone else at the table and a right to be heard with equal voice.

Today, one understanding of a "right" is a legalistic, human construct based on some sense of moral privilege—of social entitlement. Although a right in a democratic system of government is created by people and defined and guaranteed by law, access to a right may not be equally distributed across society. Conversely, a right does not apply to any person outside the select group unless that group purposely confers such a right on a specifically recognized individual, such as the disenfranchised.

In a true democracy, the whole protects all of its parts, and the parts give obedience to the will of the whole. Ostensibly, therefore, a right in democracy gives everyone equality by sanctifying and impartially protecting certain socially acceptable behaviors while controlling unsanctioned ones. There is, however, a price exacted for having rights, even in a true democracy.

Rights have responsibilities attached to them. Thus, whenever a law is passed to protect the rights of the majority from the transgressions of the minority, everyone pays the same price—some loss of freedom of choice, of flexibility—because every law so passed is restrictive to everyone. Put succinctly, we give up personal freedoms in order to gain personal rights.

The problem is that rights, as granted by humans to one another in daily life, including in the United States, are based on access—not equality. Access is determined by some notion that one race, color, creed, gender, or age is superior to another, which means that differences and similarities are based on our subjective judgments about whatever those appearances are. In American society, for example, men are still judged more capable than women in most kinds of work because society has placed more value on certain kinds of products, that is, those demanding such masculine attributes as linear thinking and physical strength, as opposed to those demanding such feminine attributes as relationship and physical gentleness.

With notable exceptions, the stereotype holds that perceived differences in outer (superficial) values become social judgments about the inherent (real) values of individual human beings. Superficial characteristics are thus translated into special rights, privileges, or entitlements simply because the individuals involved are different in some aspects and either perform certain actions differently or perform different actions. The greater the difference a person perceives between another person and him- or herself (such as a millionaire versus a homeless beggar), the more likely they are to make black-and-white judgments about that person's real value, as expressed through their notion of that person's rights.

Stated in a generic sense, such judgments are made against the standard I (in the generic sense) use to measure how everything around me fits into my comfort zone. I thus judge people as good or bad, depending on how they conform to my standard of acceptability, a standard taught and reinforced by my parents and later by my peers and teachers: social conditioning. Such judgments are erroneous, however, because all I can ever judge is appearances. In addition, my standard is correct for me only; it is not validly imposed on anyone else. Nevertheless, I use socially constructed, hierarchical couplets of extrinsic differences (white male versus white female, white male versus black male, human versus Nature) as a basis for judging the equality of such things as one race versus another, men versus women, secular versus spiritual, right versus wrong, good versus evil, and so on.

The most extreme example of personal judgment is the use of superficial differences to justify a social end. One group of people thus declares itself superior to another group because it wants what the other group has. The "superior" group tells those in the "inferior" group that they have no rights, and through this denial of rights justifies its abuse of fellow human beings.

When, for instance, the invading Spanish conquered the Pueblo Indians, they could not accept, let alone acknowledge, that they and the Pueblos were equally human. Had they acknowledged that truth, they could never have justified the wholesale murder of the Indians and theft of their land. In turn, when the invading Anglos conquered the Spanish, they could not accept, let alone acknowledge, that they and the Spanish were equally human. Had they acknowledged that truth, they could never have justified the wholesale murder of the Spanish and theft of their land. Modern conquests continue, even if economic, and so does the cycle.

The same principle holds for the indigenous peoples of the South American tropical forests. If the cattle barons ever admitted that the indigenous peoples living in the forests were their equals, they could not clearcut and burn the forests to gain pasture for their beef herds. In creating the pastures, the cattle barons destroyed an ecosystem and stripped the indigenous peoples not only of their current livelihood but also of their future options and those of their children. If the cattle barons were to admit that the indigenous peoples are in every way their equals, then they would have to *treat* them as their equals, and that, in turn, means sharing control of their mutual social destiny.

It is not a question of who is better than whom. Rather, it is a two-fold question of who is more afraid of whom and who can never have enough of what the other person has. It is also a question of who has internalized all the assumed differences and therefore perceives another human being as an undesirable entity. It is a question of who is so afraid of losing control of their perceived rights that they will do anything to keep control, regardless of social and environmental consequences. In the end, therefore, it becomes a question of the equality of one another's differences as human beings.

The Equality of Differences

Notions of superiority and inferiority are based on personal, familial, and societal judgments about the intrinsic values of extrinsic differences. To illustrate, consider two questions about garbage collectors and medical doctors: Is collecting garbage, as a social service, of equal value to that of treating sick people? Is the social stature of a garbage collector equal to that of a medical doctor?

Most people, in Chris's experience, seem to think that the service performed by medical doctors is of greater social value than that performed by garbage collectors, and that doctors not only enjoy but also deserve a higher social status than garbage collectors deserve. But, when judging garbage collectors versus medical doctors, most people focus on differences and fail to take similarities into account, one of which is that both occupations help maintain a healthy environment for people. Both occupations also rely on each other's services. In reality, doctors probably rely on garbage collectors—although unconsciously—more than garbage collectors rely on doctors.

How much more difficult would be the doctor's task if garbage collectors allowed human refuse to accumulate around houses and in streets? The outcome could be an epidemic of bubonic plague, a disease carried by rats that proliferate in human garbage and whose fleas transmit the disease. Once plague bacteria began spreading, doctors not only would have to marshal their numbers to treat the sick but also, perhaps unconsciously, would rely more directly on the personnel of sanitation districts to control the rats.

Garbage collectors serve society before the fact at a fundamental, collective level. Medical doctors tend to serve society after the fact, after someone is ill, one individual at a time. We therefore become personally acquainted with our doctors but not usually with our garbage collectors. Such a personal acquaintance greatly increases the value people tend to attribute to an individual's job because they have a more intimate sense of the person's intrinsic value, as well as greater knowledge of how that person's profession contributes to society and their own welfare.

Nevertheless, garbage collectors are as vital to human health as medical doctors, only in a different way. Why, therefore, are they not afforded equal status in society? Perhaps the reason is that they do not need to attend school for 7 to 8 years to become sufficiently trained to collect garbage and therefore do not have a socially coveted title before their names.

Perhaps it is because few people see them at work and therefore do not ponder the value of the service they perform. Perhaps it is because we do not go to garbage collectors to make us feel better when we are ill and thus do not form a personal relationship with them, as we do with our doctors. (Have you, for instance, ever thanked your garbage collector for taking away your trash, as you have thanked your doctor for making you feel better?) Perhaps it is because, compared to medical doctors, garbage collectors do not make nearly as much money, so we deem them less successful in a society, where monetary affluence is the measure of success and social status. Perhaps it is because they may be filthy and stink when they get off work, instead of being clean and well groomed.

These notions notwithstanding, garbage collectors and medical doctors are of equal value professionally, albeit different in how they serve the health needs of society. Further, their services are not only vital but also complementary in that they accomplish far more together than either could possibly accomplish alone.

There may be a number of judgmental reasons for these discrepancies in social stature, but none of them can be applied in the context of the real value of each person. An appropriate analogy might be the spokes of a wheel. Each spoke is slightly different and seemingly independent of the others; yet, each is equal in its importance to the functioning of the wheel. Each spoke is connected at the center of the wheel and at the outer rim. Leave out one spoke and the strength and function of the wheel is to that extent diminished, although the effect might not become immediately apparent.

Each person has a gift to give, and each gift is unique to that person and critical to the whole of society. All gifts are equal *and* different. What is true for individual human beings is true for cultures and societies because each is equal in its service to the Earth. Each life, each culture, each society is equally important to the evolutionary success of our planet, whether we understand it or not. Each also has its own excellence and cannot be compared to any other. All differences among people, cultures, and societies are just that—differences. The hierarchies or judgmental levels of value are human constructs that have nothing to do with reality. Every life, culture, and society is a practice in evolution, and each is equal before the impartial Law of Cosmic Unification and its subordinate, but inviolable, biophysical principles (discussed in Chapter 5).

We must therefore discard our view of Earth as a battlefield of subjective competition, where our "human superiority" reigns over that of Nature and where "my superiority" reigns over yours. We will all be better off if we, instead, consider Earth in terms of complementary efforts in which all gifts are equal, and each in its own way is important to the health and well-being of the whole living system. Why? This is so because life demands struggle and tenacity, with life continually fitting and refitting each living thing to its function. Complementary efforts, such as those of garbage collectors and medical doctors, imply equality among people, and human equality brings us to the notion of the inalienable right of all people to environmental justice.

Environmental Justice Is Predicated on Human Equality

The concept of environmental justice, from the human point of view, asserts that we owe something to every person sharing the planet with us, both those present and those yet unborn. But, you may ask, what exactly do we have to give? The only things we have of value are the humility, love, trust, respect, and wisdom gleaned from our life experiences, which are embodied in the ramifications of each and every option we pass forward. And, it is exactly because options encompass all we have to give those living today and all the children of tomorrow, and beyond that environmental justice, as a concrete, social practice, must, of necessity, be examined within the context of human equality.

A wonderful example of perceived inequality among humans took place some years ago in Chris's hometown of Corvallis, Oregon; it involved people living in the city versus those living in the country. A farmer had been arrested and fined for throwing garbage on somebody's lawn in town. But, as circumstances played out, it became apparent that, despite the US Constitution, some people are a lot more equal than others are. The episode went as follows:

Joe City, who lived in Corvallis, took his garbage out to the country and dumped it on Bill Rural's property near Bill's house. Although Bill did not see Joe dump his garbage, he found an invoice in the garbage with Joe's name and address on it. So, Bill picked up all of Joe's garbage and drove into Corvallis, where he gave it back to Joe by dumping it on Joe's front lawn. Joe went to the police and complained.

Even though Bill said that the people of Corvallis were continually dumping their unwanted garbage on his land, and that, in this case, he was sure it was Joe's garbage because of the invoice, Joe had legal standing and Bill did not. Bill was arrested and fined, but *nothing* happened to Joe—something that sent a clear message of inequality to Joe, to Bill, and to everyone else. The message: It is okay for city folks to dump their garbage with legal impunity on the property of rural folks, *but not vice versa.*

Let us look at this scenario another way. Rural people, who value clean air and quality water, have a right to enjoy these amenities, especially when they purposefully live "in the middle of nowhere." But, bureaucrats hundreds of miles away give cities and industries the right to pollute air and water because of economic and political pressures. They do this despite the fact that such pollution fouls the air and contaminates the water rural people have no choice but to use.

Human inequality has to do with fear and its companion, control. The person who harbors the most fear also harbors the greatest need to be in control of their external, materialistic environment. This need to be in control is always fed by the need for the "inequality of others," who are often demonized in one way or another in order to steal with impunity their "personal rights"—by making them of lesser stature—for self-interested gain.

Inequality, which translates into injustice, carries over into every institution in our land, but it is perhaps clearest in those agencies whose missions are to uphold and fulfill the legal mandates of protecting environmental quality for all citizens, present and future. It is seen everywhere in the appalling lack of evenhandedness and in the bending of people within the agencies (including the Congress of the United States) to the political pressure of special interest groups at the expense of all generations, especially those of the future who reap the ever-increasing social-environmental consequences with no recourse.

There have been times, however, when equality and justice counted for something, as Thucydides said of the Athenian code, "Praise is due to all who ... respect justice more than their position compels them to do." And, more recently, the founding fathers of the United States did their best, through the Declaration of Independence, the Constitution, and the Bill of Rights, to instill equality and justice in the new nation—but still only for white (European) men.

Nevertheless, agencies and individuals responsible for the welfare of our nation's natural resources have functions prescribed by law, but not necessarily specified by law. Legislative bodies therefore permit a wide range of administrative discretion. The system lacks a guiding precept for "public service," one that in fact means serving the whole of the public with the impartiality of justice for the common good of all generations.

These agencies and individuals are most often under great political pressure from special interest groups. This pressure, exerted through elected politicians, results in dedicated public servants being captives not only of the traditions of their organization but also of the fears and political weaknesses of their superiors. Therefore, true public servants are subjected to conflicting demands and receive no assurance or ethical governance, despite their dedication to such governance. As a result, our system of caring for the nation's natural resources has neither an ethical standard, or "ethos," nor a sense of social-environmental justice within society itself or toward the environment that nurtures and sustains society.[3]

Ethos, a Greek word meaning "character" or "tone," is best thought of as a set of guiding beliefs that, as mentioned previously, is neither clearly articulated nor ubiquitous in most state and federal land management agencies. In phrasing this guiding direction, a distinction needs to be made between ethos and policy. Policy, written in explicit terms, can be in the form of an order—the letter of the law. Ethos, on the other hand, is implicit and includes a guiding set of human values—the heart of the law—that is understood but cannot be easily written out. Yet, ethos can be translated into policy should one wish to do so.

Instead of a clearly articulated ethos translated into a policy of environmental justice and human equality, however, our society is both arrogant and greedy. Arrogance arises from the ignorance that assumes present knowledge is both correct and unchanging. Greed, which fosters hoarding, is born out of the fear of loss, the fear of never having enough in the material sense. And, both are justified—even hallowed—in the economic theory that underlies our capitalistic way of doing business, which is but a reflection of our social psyche, out of which arise competition and its attendant conflicts.

Perceived Resource Scarcity Accentuates Environmental Conflict

Nature affixes no human-derived values to either its components or the interactive whole. Its intrinsic value is sufficient unto itself. Humanity, on the other hand, puts its own values on some portions of the environment at the expense of others. People then compete for those items of Nature in which they find common value and squander those in which they do not.

That said, environmental conflict is born out of a perceived threat to a person's "right of survival," however that is defined. The perceived security of our right to survive is weighed against the number of choices we think are available to us, as individuals, and our ability to control our choices. Thus, as long as one party in a conflict thinks it can win agreement with its stance, which means to defend its perceived choices, that party will neither compromise nor change its position.

Perceived choices are ultimately affected by the real supply and demand for natural resources, the source of energy required in one form or another by all life. Perhaps, with this in mind, former Soviet leader Mikhail Gorbachev asked, "If we're going to protect the planet's ecology, we're going to need to find alternatives to the consumerist dream that is attracting the world. Otherwise, how will we conserve our resources, and how will we avoid setting people against each other when resources are depleted?"[4]

The greater the supply of a particular resource, the greater the freedom of choice an individual has with respect to that resource. Conversely, the smaller the supply and the narrower the range of choices available often incite people to steal choices from one another to augment their own sense of insufficiency. And, scarcity, real or perceived, is the breeding ground of environmental injustice, which rears its ugly head each time someone steals from another rather than taking responsibility for their own behavior and sharing equally. Here, a paradox arises like the phoenix out of the ashes of conflict: The people who have the most, often want the most, and thus—perceiving a sense of scarcity—are usually the most competitive and combative in securing what they need and want.

An excellent example of environmental injustice occurred in 1991, when the people of Las Vegas and Clark County (in southern Nevada) attempted to take their neighbors' water against their neighbors' will. Las Vegas not only is built in a very fragile desert, where no city should exist, but also is made up of many people who squander more water than people anywhere I (Chris) have ever been—and they can least afford it. Nevertheless, during the 2 years that I and my wife, Zane, lived in Las Vegas, the gutters of the streets ran almost every morning with great streams of water, some of which extended for a quarter of a mile or more.

The squandered water came from uncontrolled irrigation used solely to keep household and corporate lawns green. In addition, water was squandered on numerous artificial lakes and ponds and countless open swimming pools. Rather than conserving its limited supply, the city and county coveted the water of their northern neighbors and tried to figure out ways to get it, very much against the wishes and the will of those to whom it "belonged."

In short, rather than accepting the limitations of the desert in which they chose to live, the people of Las Vegas and Clark County were trying to usurp the choices of others. If they succeeded, those who used water wisely and therefore had a greater number of options would be unjustly penalized for their thrift. The people of Las Vegas and Clark County, on the other hand, could continue squandering water with impunity by taking from others, thus avoiding personal restraint and accountability for their extravagant use of water. (Read "Dissonance within Duplicity: Sustainability Programs at the Southern Nevada Water Authority, a Case Study," by Jessica K. La Porte, for a thorough discussion of this theft of water.[5])

The variety of available choices dictates the amount of control one feels they have. This consequently affects their sense of security about their survival. What would happen should you perceive your array of choices as fading or when they have suddenly been ripped away, as would happen if the people of Las Vegas and Clark County ever manage to permanently steal their neighbors' water? Have you ever been told that you can no longer do something you have always done and therefore have taken doing it for granted? How did you feel?

How would you feel if you were suddenly plucked from whatever you are doing without warning and for no apparent reason, thrown into prison without explanation or recourse, and held indefinitely against your will? Are there such innocent people behind bars now? The answer is yes!

How would you feel if you were jerked out of the only life you know, smuggled into an alien country, and sold into the bonds of slavery, never again to see anyone or anything you knew or to enjoy the rights you once had as a citizen of the United States? This is not a far-fetched scenario; slavery is very much alive in the world today—even here, in the United States.[6]

How would you feel, as an average citizen with no political desires, if civil war (such as that which is occurring in Iraq, Afghanistan, and parts of Africa) erupted suddenly all around you, and you had nowhere to go while your family, home, and town were being blown apart? I (Chris) have had a very small taste of this powerless feeling when held at gunpoint on at least one occasion in Egypt while the country was under Nasser's dictatorial fist and again by the Communist Chinese in Nepal, where I was conducting medical field research.

We ask these questions, even though they are not specifically resource oriented, in the hope that you can imagine how you would feel deep within if you were suddenly to lose your sense of safety and well-being—your sense of choice. We ask because we are convinced that conflicts arise from a deep, albeit usually unconscious, sense of potential loss: a chronic or acute fear of the future based on an unconsciously socialized disaster mentality.

For example, the functional premise of the global stock market is based on the fear of loss. That is why people buy stocks when the market is high and sell them when the market is low. They are afraid of missing an opportunity to make more money when the market is high, hoping it will continue to climb. When, however, the market value falls, as it is destined to do, they panic and sell in fear of losing what capital they have in whatever stocks are declining in value. They lose in both cases, however. On the other hand, they would gain at both ends if they sold when the market was high and bought when it is low, but first they would have to overcome their fear of loss.

Resource Overexploitation and the Fear of Imaginary Loss

According to a song popular some years ago, "freedom's just another word for nothing left to lose,"[7] which in a peculiar way speaks of an apparent human truth. When I (in the generic sense) am unconscious of a material value, I am free of its psychological grip. However, the instant I perceive a material value and anticipate possible material gain, I also perceive the psychological pain of potential loss, usually to someone else.

The larger and more immediate the prospects for material gain, the greater the political power used to ensure and expedite exploitation because not to exploit is perceived as losing an opportunity to someone else. And, it is this notion of loss that people fight so hard to avoid. In this sense, it is more appropriate to think of resources managing humans rather than of humans managing resources.[8]

For example, the administration of President Donald Trump has withdrawn the United States from the Paris Agreement on climate, all to protect the economic interests of the coal, oil, and gas corporations, to the social-environmental detriment of all generations and the planet as a whole—an action in which history is being repeated once again.[9]

Historically, any newly identified resource is inevitably overexploited, often to the point of collapse or extinction. Its overexploitation is based, first, on the perceived rights or entitlement of the exploiter to obtain their share before someone else does and, second, on the exploiter's right or entitlement to protect their economic investment. There is more to it than this, however, because the concept of a healthy capitalistic system is one that is ever growing, ever expanding, but such a system is not biologically sustainable.[10]

With renewable natural resources, such nonsustainable exploitation is a "ratchet effect," where to ratchet means to constantly, albeit unevenly, increase the rate of exploitation of a resource.[11] The ratchet effect works as follows: During periods of relative economic stability, the rate of harvest of a given renewable resource, say timber or salmon, tends to stabilize at a level that economic theory predicts can be sustained through some scale of time. Such levels, however, are almost always excessive because economists take existing unknown and unpredictable ecological variables and convert them, in theory at least, into known and predictable economic constant values in order to better calculate the expected return on a given monetary investment from a sustained harvest.

Then comes a sequence of good years in the market, in the availability of the resource, or both, and additional capital investments are encouraged in harvesting and processing because competitive economic growth is the root of capitalism. When conditions return to normal or even below normal, however, the industry, having overinvested, appeals to the government for help because substantial economic capital, and often many jobs, is at stake. The government typically responds with direct or indirect subsidies, which only encourage continual overharvesting.

The ratchet effect is thus caused by unrestrained economic investment to increase short-term yields in good times and strong opposition to losing those yields in bad economic times. This opposition to losing yields means there is great resistance to using a resource in a biologically sustainable manner because there is no predictability in yields and no guarantee of yield increases in the foreseeable future. In addition, our linear economic models of ever-increasing yield are built on the assumption that we can in fact have an economically *sustained* yield. This contrived concept fails in the face of the biological *sustainability* of a yield.

Then, because there is no mechanism in linear, economic models of ever-increasing yield that allows for the uncertainties of ecological cycles and variability or for the inevitable decreases in yield during downtimes in the market, the long-term outcome is a heavily subsidized industry. Such an industry continually overexploits the resource (e.g., water available for human use, in the case of agriculture) on an artificially created, sustained-yield basis that is not biophysically sustainable.

When the notion of sustainability arises in a conflict, the parties marshal all scientific data favorable to their respective sides as "good" science and discount all unfavorable data as "bad" science. Environmental conflict is thus the stage on which science is politicized, largely obfuscating its service to society.

Because the availability of choices dictates the amount of control we humans feel we have with respect to our sense of security, a potential loss of money is the breeding ground for environmental injustice. This is the kind of environmental injustice in which the present generation steals from all future generations by overexploiting a resource rather than facing the uncertainty of giving up potential income.

There are important lessons in all of this for anyone mediating environmental conflicts. First, history suggests that a biologically sustainable use of any resource has never been achieved without first overexploiting it, despite historical warnings and contemporary data. If history is correct, resource problems are not environmental problems but rather human ones that we have created many times, in many places, under a wide variety of social, political, and economic systems.

Second, the fundamental issues involving resources, the environment, and people are complex and process driven. The integrated knowledge of multiple disciplines is required to understand them. These underlying complexities of the physical and biological systems preclude a simplistic approach to both management and conflict resolution. In addition, the wide, natural variability and the compounding, cumulative influence of continual human activity mask the results of overexploitation until they are severe, which progressively increases their irreversible consequences.

Consider, for example, that on March 29, 2017, newly elected US President Donald Trump signed an executive order sweeping away Obama era policies on climate change:

> US President Donald Trump has signed an executive order undoing a slew of Obama-era climate change regulations that his administration says are hobbling oil drillers and coal miners. . . .
>
> The decree's main target is former President Barack Obama's Clean Power Plan, which required states to slash carbon emissions from power plants—a critical element in helping the United States meet its commitments to a global climate change accord reached by nearly 200 countries in Paris in 2015.
>
> The so-called Energy Independence order also reverses a ban on coal leasing on federal lands, undoes rules to curb methane emissions from oil and gas production, and reduces the weight of climate change and carbon emissions in policy and infrastructure-permitting decisions.
>
> "I am taking historic steps to lift restrictions on American energy, to reverse government intrusion, and to cancel job-killing regulations," Mr [sic] Trump said at the Environmental Protection Agency headquarters, speaking on a stage lined with coal miners.

> The wide-ranging order is the boldest yet in Mr Trumps [sic] broader push to cut environmental regulation to revive the drilling and mining industries, a promise he made repeatedly during the presidential campaign.
> . . .
> "I cannot tell you how many jobs the executive order is going to create but I can tell you that it provides confidence in this administration's commitment to the coal industry," Kentucky Coal Association president Tyler White told reporters.[12]

Third, as long as the uncertainty of continual change is considered a condition to be avoided, nothing will be resolved. However, once the uncertainty of change is accepted as an inevitable, open-ended, novel, creative life process, most decision-making is simply common sense. For example, common sense dictates that one would favor actions having the greatest potential for sustainability, as opposed to those with little or none. Such sustainability can be ascertained by monitoring results and modifying actions and policy accordingly. It must be understood, however, that nothing in the universe is reversible because the process of change is a biophysical constant and thus always novel in its outcomes, as discussed in the next chapter.

Fourth, the seed of every conflict is based on a sense of loss—which usually translates into a lifelong fear of loss in some degree—that originates in childhood, as our social conditioning and its entrenched perspectives, from the expressed concerns of our parents, teachers, news media, and peers. Consequently, our fears are ultimately interpreted, and continually reinforced, as a threat to our personal survival.

Conflict Is a Mistake

The result of many childhood lessons is the perceived need for control, and anything that threatens our control is an enemy onto whom or which we can project blame for our fears and thereby justify them. But, who or what is the enemy? An enemy is one seeking to injure, overthrow, or confound an opponent; it is something harmful or deadly. Of course, *I* (in the generic sense) am not "the enemy" because I am convinced that *my* position and *my* values are the *right* ones, and everyone knows that "the enemy" is wrong. That is what we are taught. That is the eternal verity around which conflict rallies.

The problem is that when all sides feel justified in their points of view, there is little understanding that an enemy is anyone or anything opposing that view. Opposition to that view is thus perceived as a threat to survival, however it is defined. Herein lies the great irony: Most environmental conflicts are the spawn of misunderstandings, miscommunication, and misperceptions in one way or another. Conflict is thus often a mistake, a misjudgment of appearances, or an assumption that is avoidable because it is only one choice—a *reaction*—to a given circumstance. Other *responses* are available should a person examine them and choose to accept an alternative.

Consider war, the ultimate destructive conflict, both socially and environmentally. War, as is all human conflict, is based on the personalities of the people involved (in this case, the leaders) and their common feelings about fear and enemies.

Once one side or the other perceives a threat to its survival, the single most important precipitating factor in the outbreak of war is, in fact, misperception, which manifests itself in a leader's self-image and the leader's view of the adversary's character, of the adversary's intentions, capabilities, and military power. Once misperception is in play, miscommunication closes in and joins hands with misjudgment to foster a distorted view of the adversary's character, which helps to precipitate the conflict.

If a leader on the brink of war believes the adversary will attack, the chances of war are fairly high. If both leaders share this perception about each other's intent, war becomes a virtual certainty.

But, it is a leader's misperception of the adversary's power, and willingness to use it, that is perhaps the quintessential cause of war. It is vital to remember, however, that it is not the actual distribution of power that precipitates a war; what precipitates war is the way in which a leader *thinks* the power is distributed. Thus, on the eve of each war, at least one leader, through miscommunication, misperceives and thus misjudges the other's available power and willingness to use it. In this sense, the beginning of each war is a folly of misperception. The war itself then slowly and agonizingly teaches people the terribly high cost of destructive conflict.[13]

Conflict Is Usually Based on the Misjudgment of Appearances

The lesson war has to teach us is that conflict of any kind is a cycle of attack and defense based on the fear of uncertainties and unknowns, which usually results in the *misjudgment of appearances*. Appearance is an outward aspect of something that comes into view; judgment is the process of forming an opinion or evaluation of the appearance by discerning and comparing something believed or asserted. Therefore, those whom we (in the generic sense) define as enemies are those onto whom we affix blame for our perceived sense of insecurity, a perceived threat to our own survival.

Our judgments are almost always incorrect, however, because things are seldom as they appear because appearance is external. If, therefore, we could but understand the inner motive of our "enemy," we would likely find a mirror reflection of our fears for our own survival. In that reflection, we would find not only that we were mistaken about our enemy's motives but also that we had made an incorrect judgment of our enemy's character or ability based on inadequate knowledge.

Well, you might ask, if we are not one another's enemies, what is the enemy? Of what are we really afraid? We are largely afraid of change, loss of something we value through circumstances we perceive as a threat to our sense of survival because we cannot control them.

Yet, almost every circumstance we encounter in some way evokes an unanticipated change in our participation with life. In turn, each change we are obliged to make is a compromise in our sense of control, a frightening condition of life to most people in our increasingly complex, technological world.

Control, often used as a synonym for a sense of personal power, is an interesting phenomenon in life. We pay dearly for control, but regardless of the price, there are limitations. We cannot, for example, control the wind, but we can trim the sails on our boat. The wind is the circumstance beyond our control, but by trimming the sails, we can choose how our boat—and thus we—respond to the wind. And, in our response, we are in control of ourselves, which de facto controls the outcome of the circumstance.

Have you ever had a "bad" day, a day when nothing seemed to go "right," according to your desires, so you felt out of sorts, and every little thing that could go awry delighted in doing so, which unduly annoyed you, causing you to say, "I'm at my wit's end! If one more thing happens I'm going to explode!" Because you feel out of sorts or not at peace with yourself, you are therefore compelled to control the environment around you. Your inner sense of survival is shaky, and the only way you can ride out the inner storm is to have outer calm.

Now, think of a "good" day, a day when everything went right. You felt in tune with the world, and you had a feeling of inner control and peace—a day when you said, "Everything I touch turns to gold!" On that day, the external things that still delighted in going awry did so, but they did not bother you.

Was it, in fact, that the day was *good* or *bad*? Or, was it how you felt about yourself and your sense of survival on that given day? The difference between the 2 days was simply how you *responded* to a given circumstance on the good day or *reacted* to the same circumstance on the bad day, based on how you were feeling about yourself.

That no one can control circumstances is a given, although people continually try, which results in either inner or outer conflict of some magnitude. But, we can control whether we react or respond to a circumstance, and therein lies both our problem and its potential resolution.

Our inability to control external circumstances in any meaningful way translates into fear of change because every circumstance causes something to shift, and every alteration is novel and irreversible. We, however, tend to focus on a potentially negative outcome—be it relatively minor (such as finding a tiny nick in the fender of our new car) or catastrophic (such as floodwaters pouring through our home). We thus perceive change as a loss of control that threatens our survival in one way or another. We therefore want to control circumstances whenever we can, so that other people—our perceived enemies—will have to risk change, but we will not.

There are no true enemies, only people frightened of the same kinds of things that we are afraid of. We thus mistakenly reject these people as enemies, a judgment we use to justify our side of a conflict in the "war" of survival, the war to "control" circumstances beyond our control.

When we focus our attention on human enemies, what we are really focusing on is not the outer enemy, but rather our fear of the unknown, of losing our sense of control based on the familiar, which dehumanizes our soul. Conflict is thus our attempt to place our problem onto someone or something else, to move away from our fear, away from something we do not want to have happen.

Conflict Is Merely a Product of Our Mutual Choices

To help us understand the mutuality of choices, we use the dynamics of the hydrologic cycle, which is the circulation of water in Earth's system: from the ocean to the land (Figure 2.1). The hydrological dynamics encompasses the principle pathways that water takes (precipitation, infiltration, overland flow, percolation, stream flow, groundwater flow, evaporation, and so on)

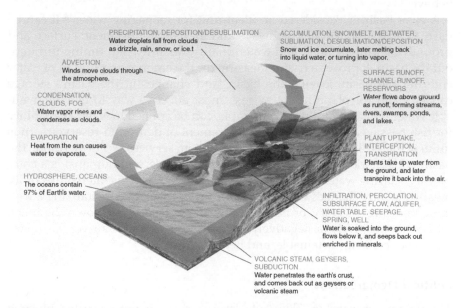

FIGURE 2.1
Diagram of the water cycle, showing how water is stored and transported across the planet. (From Ehud Tal, Wikimedia, 2016.)

and places wherein water is stored at various scales and for lengths of time (in the atmosphere, glaciers, soil, biomass, groundwater, rivers, lakes, and oceans).[14] This circulatory aspect means that water can be studied from any part of the cycle and thus accessed at multiple scales across Earth's entire biophysical system or within a given landscape. Examining the pathways and storage areas of water at the global extent is informative and wise, while at the local and regional scales, it helps with forecasting the availability of water for human use,[15] whether in a city, county, province, or even a particular river basin.

River Basin

A river basin is synonymous with a drainage basin, watershed, or catchment. It encompasses all the water in a landscape that is connected to a single river system. Precipitation (rain, hail, sleet, and snow) moves through (and over) land to a body of water, such as a lake or ocean. The highest points (ridges) within the landscape set the divide (Nature's boundaries), separating one basin from another, thereby causing precipitation to fall into different river systems at different times and concentrations. Movement of water through the basin (including its connectivity with groundwater) is governed by climate, geology, topography, soil, vegetation, and human activity. When these variables, along with basin size and stream characteristics (stream order hierarchy and length), are similar among basins, the more alike basin waters behave.[15]

A river basin is considered an ideal unit for managing water.[16] It can be used to learn about water pathways and its rate of movement; surface and groundwater interactions; and water quality, quantity, and timing throughout the basin. It proves a useful measure for the amount and timing of water entering the basin, leaving the basin, and changes in the basin's storage capacity and for planning the wisest, most effective way to use the available water. This accounting, for example, can be applied to water availability for fish passage, irrigation, and the placement of dams and reservoirs, thus influencing basin "commodities and ecosystem services."[17] Furthermore, the basin-wide biotic ecosystem (fungi, plants, animals, and humans) is a community, sharing waters within its natural boundaries. This interconnectivity means that any changes within the water catchment, like revitalizing a wetland, building a factory, or flushing pharmaceutical drugs down the toilet, will either positively or negatively impact the capacity for a river basin system to be resilient, sustainable, and healthy.

Political Demarcations

The volume of water or "flow" of Nature's rivers is not restricted by such invisible, artificial constraints as legal boundaries (property lines) or political boundaries (township, provincial, and state demarcations). As surface

Transboundary river basins of the world

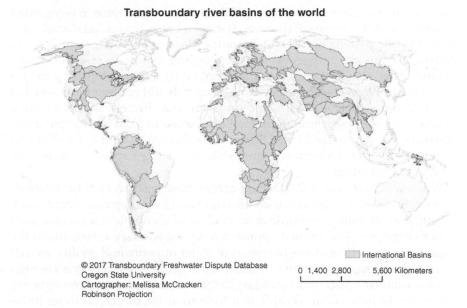

International Basins

© 2017 Transboundary Freshwater Dispute Database
Oregon State University
Cartographer: Melissa McCracken
Robinson Projection

0 1,400 2,800   5,600 Kilometers

FIGURE 2.2
A map of the world's transboundary river basins. (Transboundary Freshwater Dispute Database [TFDD], 2017. Product of the Transboundary Freshwater Dispute Database, College of Earth, Ocean, and Atmospheric Sciences, Oregon State University. Additional information about the TFDD can be found at http://transboundarywaters.science.oregonstate.edu.)

and groundwater crosses international borders, additional layers of complexity are added to manage these transboundary waters. In fact, worldwide, over 300 river basins (Figure 2.2), and an even larger number of aquifers, intersect a country's boundary. And, approximately, 40 percent of humans reside in transboundary river basins.[18] While most of these basins are shared by two or three countries, some span a larger number of nations. As many as 7 countries share the Jordan Watershed (Middle East); the Amazon Basin (South America) comprises 8 nations; 10 countries are part of the basin of the Nile River (Africa); and 19 countries share the Danube River Basin (Europe; Wolf et al., 2005).[19] At the state and provincial levels, the Columbia River Basin is made up of the province of British Columbia (Canada); the states of Idaho, Oregon, Washington, Montana, Nevada, Utah, and Wyoming (United States); as well as 15 tribal reservations—each a sovereign entity.

Civilizations and communities develop where freshwater is readily accessible, but a nation or city thrives or declines based on the "control" of water.[20] Such attempted control impacts not only the economy and security of nations but also the health of the ecosystem and its human population.[21] As a result, when water crosses political boundaries, people, cities, and countries lay claim to it to protect their city-, county-, and country-based interests. This territorial perspective of water as a resource negates a communal or conservation approach.[22] Because a basin approach is regarded as a more holistic approach to water management at the national level, promoted by the passing of the 1991 Resource Management Act, New Zealand restructured its environmental policy according to a basin system.[23] In the United States, a hydrologic unit (a regional code) was developed by the US Geological Survey. However, a hydrologic unit should not be strictly equated to basin units because hydrologic units are not discrete basins as some basins have been *conveniently* grouped or divided to form hydrologic units. Nevertheless, within the United States, more than 20 states use a basin management approach across their entire state, including Kentucky, Ohio, New Jersey, and Washington.[24]

When water cooperation is lacking across these legal and political demarcations, there can be significant losses in terms of lives, impaired community health, poor planning, and little or no pooling of economic resources or joint water ventures.[25] This limited approach to transboundary waters limits the capacity to properly address the issues of water quantity and quality, as well as waterborne diseases. Under these circumstances, when tensions are high among water-sharing, transboundary neighbors (communities and nations), there may be little, if any, sharing of information. So, unilateral water infrastructure projects are more prevalent, and due to a lack of communication, this inversely affects neighbors.[26]

Water Conflicts and Cooperation Between Nations that Share Water Resources

With an ever-growing world population, fast approaching 9 billion by 2037, as compared to the time frame of 1960, when it was about 3 billion,[27] coupled with competing demands for water and a changing climate, the media is touting that the world is on the verge of "water wars." To determine what validity this water war argument has, Oregon State University researchers documented national water interactions among nations.

This research revealed that, globally, water can place stress on stable international bonds and strain poor relationships between neighboring countries that *share* water. Most such interactions among nations are, however, considered mildly cooperative or mildly conflictive, falling short of voluntary unification into one nation or a formal declaration of war. The most recent water war happened 4,500 years ago between city-states Umma and Lagash, now within Iraq. Since then, no nation has started a war specifically because of water.[28] In fact, it was determined that globally there are double the number of cooperative water interactions between nations (water treaties, agreements, and mild support) than there are water conflicts (hostility expressed verbally, politically, or militarily). And, when effective institutions exist, water can be a galvanizing force among nations that share water resources.[29]

Moreover, the research indicated that even the most contentious water conflicts among nations are eventually resolved. Delli Priscoli and Wolf indicated that it took 10 years for the Indus Treaty to be signed between India and Pakistan, 30 years for resolution concerning the Ganges River between India and Bangladesh, and 40 years between Jordan and Israel in the case of the Jordan River. During these decades of unresolved conflicts, there is, however, degradation of water quality, biodiversity, wetlands, and human health.[30] At the international scale, cooperative water events are most often over quantity, rather than quality, followed by economic development and joint management. In fact, 90 percent of conflicts over water are about quantity and infrastructure.[31]

Water Conflicts and Cooperation Within a Nation

Different geographic scales (local, national, and international) set the stage for different dispute dynamics. At the national level, more intense hostility is regarded as more prevalent than at the international scale.[32] Compared to international conflicts, national disputes can involve many more stakeholders. For example, stakeholders might comprise branches of federal, state, county, and city governments; indigenous peoples; city and rural residents; watershed coalitions; nongovernment organizations; agricultural and industrial representatives; logging communities; fisheries; and environmentalists. With different demands, all are vying for water resources: Some depend on historical water use or ancestral water claims, while others want to protect endangered species or instream flow or have property rights legitimized for access to water.

Research conducted by Kristel Fesler, of Oregon State University, concerning conflicts and cooperation over water resources within 4 of the 18 Oregon basins, focused on the time frame between 1990 and 2004. Her data indicate that, similar to the international level, even at the country-state scale there are more cooperative than contentious outcomes, with issues of instream flow garnering the highest number of conflicts. Furthermore, at this scale, water conflicts were exacerbated over time but did eventually lead to cooperation, whereas cooperative agreements, while tested, were more enduring and resilient. Most notable in Oregon was the fact that issues of water *quality* were more common (40 percent) than noted at the international scale (6 percent). In Oregon, 20 percent of events were about water *quantity*, compared with 45 percent at the international level.[33]

Fesler noted that, between 1990 and 2004, a series of environmental legislative decisions (such as the phosphorus ban, Oregon Plan for Salmon and Watersheds, Biological Opinions, and Local Watershed Management Plans) were being decided in Oregon. They incited both cooperative and conflictive responses among stakeholders and, consequently, had subtle differences, illustrating institutional changes at both the state and local levels.

In addition, the research showed a statistically significant relationship between the lag time from when decisions were made and community tensions, suggesting that more constructive outcomes may have resulted if greater public involvement had been more central to the decision-making process.[34]

Conflicts and Cooperation: You and Me

Conflict versus cooperation is generally regarded as an integral and inevitable part of life since we experience it at work, at home, and at play. It is experienced within national politics, as well as between and among nations (complex systems); federal environmental agencies (groups); people (interpersonally); and within ourselves, as individuals (intrapersonally). Consider that we are an integral part of complex systems and groups, and that our inner and outer selves influence and are impacted by our roles within these systems and groups. For example, as a federal employee working for an environmental agency, one may attend an international conference, thereby representing a nation (complex system), a federal entity (group), and one's privately held environmental values. Through one-on-one interactions at the conference, there is interpersonal engagement and the alignment of one's physiological, emotional, mental, and spiritual being, all of which influence one's intrapersonal experience.

So, while we refer to nations, governments, tribes, and people in the collective, keep in mind that, in fact, it is us. We are the people who represent a nation, region, community, university, neighborhood, and family. And, while we represent the mission and values of these complex human systems and groups, we nevertheless *show up* as ourselves, with our intrapersonal perspective and perceptions expressed in an interpersonal setting. The last encompasses our physiological, emotional, mental, and spiritual aspects of our being—our four worlds. When these worlds are incongruent, we often find ourselves conflicted. Aligning these four worlds is central to faith-based traditions (such as of Jewish, Buddhist, Muslim, and indigenous people of the Americas), which offer spiritual practices, such as meditation, to foster balance, harmony, and constructive outcomes.[35]

As such, our ability to effectively engage with any scale of a complex system, personally or otherwise, is ultimately governed by our inner harmony. As individuals, we are at the core of all interactions. So, when nations are negotiating over environmental issues, who is at the table? We are—you and me. Regardless of the organization or country we may represent, our way of engaging in the negotiation (or any interaction) is ultimately governed by our individual state of being—our alignment with the four worlds.

Therefore, despite the media claims that water wars are coming, we believe the answer ultimately lies within you and me. Because we are the core of human systems, both large and small, we determine our collective fate: environmentally, socially, and politically. And, if we find our current path a progressively unsustainable legacy for all generations, we have the power to change course, to choose again with greater wisdom. So, our collective, social-environmental future is governed by you and me and our alignment with the four worlds.

Discussion Questions

1. Is conflict really a choice?
2. If so, why choose to create a conflict rather than mutual cooperation? What is the thought process behind it?
3. What is meant by the "equality of differences" with respect to people? How does adherence to the equality of differences influence the resolution of a conflict?
4. What is the difference between a "sustained" harvest and a "sustainable" harvest?
5. Can we use the hydrologic cycle as a metaphor for the conflict cycle and thus study a conflict from any part of its cycle?
6. Is there a particular question you would like to ask?

Endnotes

1. Cameron La Follette and Chris Maser. *Sustainability and the Rights of Nature: An Introduction.* CRC Press, Boca Raton, FL, 2017. 418 pp.
2. See the following books by Alice Miller: (1) *The Drama of the Gifted Child: The Search for the True Self.* Basic Books, New York, 1997. 134 pp.; (2) *For Your Own Good: Hidden Cruelty in Child-rearing and the Roots of Violence.* Farrar, Straus, Giroux, New York, 1983. 284 pp.; (3) *Thou Shalt Not Be Aware.* Farrar, Straus, Giroux, New York, 1998. 330 pp.; (4) *Pictures of a Childhood.* Farrar, Straus, Giroux, New York, 1986. 178 pp.; (5) *The Untouched Key: Tracing Childhood Trauma in Creativity and Destructiveness.* Anchor Books, New York, 1990; (6) *Banished Knowledge.* Doubleday, New York, 1990. 179 pp.; and (7) *Breaking Down the Wall of Silence: How to Combat Child Labour.* Dutton Books, New York, 1991. 60 pp.
3. Robert B. Strassler. *The Landmark Thucydides: A Comprehensive Guide to the Peloponnesian War.* Simon and Schuster, New York, 1998. 752 pp.
4. Colin Greer. If We're Going to Protect the Planet's Ecology, We're Going to Need to Find Alternatives. *Parade Magazine,* January 23, 1994, pp. 4–5.
5. Jessica K. La Porte. Dissonance within Duplicity: Sustainability Programs at the Southern Nevada Water Authority, a Case Study. In: Chris Maser, *Decision Making for a Sustainable Environment: A Systemic Approach.* CRC Press, Boca Raton, FL, 2013, pp. 125–139. 304 pp.

6. Chris Maser. *The Perpetual Consequences of Fear and Violence: Rethinking the Future*. Maisonneuve Press, Washington, DC, 2004. 373 pp.
7. Kris Kristofferson. Me and Bobby McGee. http://www.bluesforpeace.com/lyrics/bobby-mcgee.htm (accessed March 13, 2010).
8. Donald Ludwig, Ray Hilborn, and Carl Walters. Uncertainty, Resource Exploitation, and Conservation: Lessons from History. *Science*, 260(1993):17–36.
9. Environmental Policy of the Donald Trump Administration. https://en.wikipedia.org/wiki/Environmental_policy_of_the_Donald_Trump_administration (accessed February 20, 2018).
10. Russ Beaton and Chris Maser. *Economics and Ecology: United for a Sustainable World*. CRC Press, Boca Raton, FL, 2012. 191 pp.
11. Ludwig et al., Uncertainty, Resource Exploitation, and Conservation; and (2) Chris Maser. *Global Imperative: Harmonizing Culture and Nature*. Stillpoint, Walpole, NH, 1992. 267 pp.
12. Donald Trump Signs Executive Order Sweeping Away Obama-Era Climate Change Policies. *ABC Net Australia*, March 29, 2017. http://www.abc.net.au/news/2017-03-29/trump-signs-executive-order-sweeping-away-obama-climate-policies/8395486 (accessed June 9, 2017).
13. The discussion of war in this paragraph is based on John G. Stoessinger. *Why Nations Go to War*. St. Martin's Press, New York, 1974. 480 pp.
14. Chris Maser. *Interactions of Land, Ocean and Humans: A Global Perspective*. CRC Press, Boca Raton, FL, 2014. 308 pp.
15. Lawrence Dingman. *Physical Hydrology*. Waveland Press, Long Grove, IL, 1994. 575 pp.
16. Claudia W. Sadoff and David Grey. Beyond the River: The Benefits of Cooperation on International Rivers. *Water Policy*, 4(2002):389–403. http://www.transboundarywaters.orst.edu/publications/publications/Sadoff%20%26%20Grey%20Beyond%20the%20River%2002.pdf (accessed February 20, 2018).
17. Christopher Lant. Watershed Governance in the United States: The Challenges Ahead. *Water Resources Update*, 126(2003):21–28. http://opensiuc.lib.siu.edu/cgi/viewcontent.cgi?article=1109&context=jcwre.pdf (accessed February 1, 2018).
18. Aaron T. Wolf (Ed.). *Hydropolitical Vulnerability and Resilience Along International Waters: Asia*. United Nations Environmental Programme, Nairobi, Kenya, 2009. 185 pp.
19. Aaron T. Wolf, Annika Kramer, Alexander Carius, and Geoffrey D. Dabelko. Managing Water Conflict and Cooperation. In: *State of the World 2005: Redefining Global Security*. The WorldWatch Institute, Washington, DC, 2005, pp. 80–207. http://www.transboundarywaters.orst.edu/publications/publications/Wolf%20-%20SOW%2005%20chap5.pdf.
20. Rajendra Pradhan and R. Ruth Meinzen-Dick. Which Rights Are Right? Water Rights, Culture, and Underlying Values. In: Peter G. Brown and Jeremy J. Schmidt, Eds., *Water Ethics: Foundational Readings for Students and Professionals*. Island Press, Washington, DC, 2010, pp. 39–58.
21. Juliet Christian-Smith and Peter H. Gleick, with Heather Cooley, and others. *A Twenty-First Century US Water Policy*. Oxford University Press, New York, 2012. 334 pp.
22. Jerome Delli Priscoli and Aaron T. Wolf. *Managing and Transforming Water Conflicts*. Cambridge University Press, New York, 2009. 382 pp.

23. Lant, Watershed Governance.
24. Environmental Protection Agency, Office of Water. A Review of Statewide Watershed Management Approaches. 2002. https://www.epa.gov/sites/production/files/2015-09/documents/review-statewide-watershed-mgmt-approaches.pdf (accessed February 1, 2018).
25. Delli Priscoli and Wolf, *Managing and Transforming*.
26. Ibid.
27. World Bank. Total Population. n.d. https://data.worldbank.org/indicator/SP.POP.TOTL (accessed February 2, 2018).
28. Wolf et al., Managing Water Conflict.
29. Ibid.
30. Delli Priscoli and Wolf, *Managing and Transforming*.
31. Wolf et al., Managing Water Conflict.
32. Ibid.
33. The paragraph is based on Kristel Fesler. 2006. Analysis of Social Interactions Concerning Oregon's Water Resources between 1990 and 2004. Unpublished data.
34. Ibid.
35. Lynette de Silva and Aaron T. Wolf. Lessons from Spiritual Practices for Water Diplomacy. In: Janos J. Bogardi, K.D.W. Bandalal, Ronald R. P. van Nooyen, and Amit Bhaduri, Eds., *Springer Handbook of Water Resources Management*. Springer-Verlag, New York, 2018.

3

Creating Meaningful Communication

Introduction

The intent of this chapter is to see how connectivity and relationships play out in the world at the system, group, interpersonal, and intrapersonal scales. With this background, ways are explored through teaching and practice that will enable society to replace *nonconstructive* patterns and incorporate more meaningful connections.

The idea that everything is interconnected is a broad statement that often stretches our minds to consider new and existing linkages, as well as theories that include disorder, uncertainty, and the notion of chaos. Such chains of reasoning include the premise that some small, localized events can have far-reaching impacts in much larger, more complex, systems, such as a typhoon being attributed to the flapping wings of a butterfly (termed *butterfly effect*) or even a seemingly local decision, such as not admitting an applicant to the Academy of Fine Arts in Vienna, Austria, which would eventually spiral into a series of events to produce an enraged dictator, like Adolf Hitler, who initiated World War II.

John Bell's theorem suggests that activities in the cosmos, even if separated by large distances, impact actions in other places.[1] If Bell's theorem holds true, there are many aspects that continue to remain unpredictable and nonlinear. Consider, however, that not all virtuous, noble, well-thought-out, and carefully implemented actions will manifest as the most environmentally effective or biophysically sustainable outcome—despite leaning toward morally appropriate end results. Nevertheless, there are principles governing the natural world that can *help* us understand our planet and how life works.[2] And, if our strategic plan fails, a *positive* intention for a fruitful outcome *may* influence the external energy field of possibilities.[3] Throughout this chapter, we explore lessons and practices whereby connections and relationships can be forged at multiple scales.

Bear in mind, as we proceed, that increases in human life expectancy, complemented by the human population boom over the last 100 years, are attributed to advances in the science of vaccine development, thereby alleviating diseases, as well as increasing farming productivity with the progressive aid of herbicides and insecticides.[4] Wackernagel et al.[5] deemed the 1980s a crucial time frame in which the human population (between 4.46 and 5.20 billion[6]) reached Earth's productive capacity. However, as the human population grew—and continues to grow—the increasing demands from the global environment include fishing, hunting, farming, logging, and the burning of fossil fuel, as well as increasing demands for housing and municipal infrastructure. After the 1980s, human demands surpassed Earth's capacity. As a result, today the global biophysical system actually requires about 1.5 years to regenerate itself following the annual demands of its human population— meaning we humans are *living beyond the Earth's productive capacity*.[7]

People are making Earth's ecosystems "sick"[8] through global pollution, and in doing so, we compromise the biosphere and Earth's ability to sustain our own existence. If the trends of humanity's current demands continue, by the end of this century we will have behaved as though we have sustainable, productive capacity of nearly three Earths available to us. The human population is not expected to plateau until the end of the century.[9] By that time frame, the United Nations projects a population of approximately 11 billion.[10] Statistician Hans Rosling thinks we can prevent the world population from reaching that number, provided global efforts to increase "child survival" and improve the standard of living for the poor are vigorously practiced.[11] Coupled with these actions, it is imperative that humans begin to act responsibly and to plan for these larger populations.

The actions mentioned speak to a disconnection between human activity and the biophysical integrity of the planet; while the human population soars, other species (salmon, spotted owls, condors, wolves, lions, whales) dwindle in number. With this in mind, we explore lessons and tools of connectivity provided at four scales of transformation: complex systems, groups, interpersonal, and intrapersonal.

Exploring Complex Systems

Perhaps Sean Carroll described this disconnect between humans and the health of the ecosystem best by stating that, "We have taken control of biology, but not of ourselves."[12] This lack of self-awareness and lack of understanding of the rippling effect of our actions has led Peter Brown to ask, "Is ours the only species fairly able to claim air, water, territory, and other necessities of life?"[13] From the regulatory perspective, such laws as the 1992 Australian Endangered Species Protection Act[14] and the US Endangered Species Act[15] have attempted to safeguard the ecology and habitat of those nonhuman species that share the planet with us, but to whom we give no political voice.

These regulations, when enforced on both private and federal lands, can appear at odds with our individual "rights" and so infringe on our chosen livelihood and impact our preferred way of life. The effect of such regulations is evident within the timber communities in the United States. The community of Sweet Home, 32 miles from the college town of Corvallis, Oregon, is a prime example. Sweet Home experienced the enforcement of the US Endangered Species Act when its community was prevented from developing and logging public land and old-growth forest wherever it occurred. This was initiated to protect the habitat of the northern spotted owls.[16]

Are these regulations put in place to shape a nation's (and global) ethical values, so that one day they may become our community and individual philosophies? Nobel Peace Prize winner Albert Schweitzer's insight on the topic of ethics came many decades earlier, in 1915, while traveling upstream on the Ogowe River in the Congo Basin. While contemplating the notion of ethics, revealed to him through a burst of insight was the expression, "reverence for life."[17]

I (Lynette) believe laws like the Endangered Species Act, when effectively employed, can be a mechanism and are a first step to protecting, honoring, and instilling a reverence for life, not only for the endangered species, but also for all life forms. Contemplate that a rich diverse plant and animal life supports truly sustainable ecosystems. Plants, after all, are vital to the circulation of water from soil into the atmosphere through evapotranspiration; through photosynthesis, they sustain atmospheric equilibrium (producing oxygen and removing carbon dioxide). Plants typically produce foods from inorganic matter, serving us with an energy source (fuel), clothing, and housing. Examine the changes that occur within the functioning of an ecosystem when a single species of plant becomes extinct. Consider that the animal world and humans depend on these systems for food supply and nourishment, and that a quarter of all prescribed medications in the United States alone contain plant-based extracts.[18]

To illustrate, consider the far-reaching effects in the African Serengeti with elimination of the rinderpest, a viral disease that is fatal to cows, cape buffalo, and wildebeest. With its eradication, there was a surge in the wildebeest population (Figure 3.1). This increase gave rise to more wildebeest predators (lions and hyenas), and, because the wildebeest thrives on short grass, there were fewer fires and more opportunity for trees to grow. Tree growth spurred an increase in the giraffe population, which feeds on these trees. As such, the wildebeest is considered a vital species to the regulation of the Serengeti ecosystem.[19]

The web of connectivity is complex. However, there are inviolable biophysical principles (Biophysical Principles 1, 7, and 9 in Chapter 5) that govern energy and life. When these are *not* adhered to, they have detrimental consequences for *our* planet.[20] So, understanding these interrelationships and honoring the *crucial* feedback loops through which the systems operate can provide us with knowledge about the possibility of more constructive human behavior and thus more effective outcomes.[21]

Systems, whether a computer system, the human circulatory system, an ecosystem, or a river basin, comprise a systemic unit of various scales that structurally support one another, each made up of integral and interrelated components that operate (behave) in a particular manner to a specific end/ purpose. For example, the human circulatory system comprises organs that

FIGURE 3.1
Wildebeest in Tanzania. (Photograph by Lynette de Silva.)

work together to sustain the vital functioning of the human body: transporting nutrients, wastes, and gases and regulating body temperature and hormones.

Systems thinking recognizes the delicate and complex framework of systems governed by biophysical principles of which we may be oblivious, and that any action we take will generate a systemic response.[22] Systems thinking allows for a thoughtful, comprehensive, and ordered approach to identifying all (or most) of the components within a unit, observing and understanding the interconnections (feedback loops) among the components in order to examine challenges presented by a system before acting to mitigate a perceived problem.[23] Systems thinking provides tools that help us understand the world through these relationships and interactions and can thus form the basis for shared beliefs about reality.[24] Furthermore, because systems are dynamic, all assumptions are limited to a specific time frame or set of variables and can include unexpected associations among components.[25]

Capturing the workings of an environmental system (systems thinking) requires integrating multistakeholder involvement because the complexity associated with the management of environmental resources requires contributions from the scientific community, government and nongovernment entities, agricultural communities, industry and businesses, policymakers and lawyers, economists, ecosystem specialists, and those representing societal interests. In certain areas of the world, this also includes the participation of religious and spiritual leaders. In addition, keep in mind that multifaceted mechanisms can be continual and unpredictable.[26]

Characteristic aspects of systems thinking include envisioning a system as changing in two ways: changing through its own evolutional process and performing a function that leads to a change (transformation). Systems thinking accounts for interruptions within a system that can enhance or decrease the original behavior (feedback loops) and in which the system is considered in its entirety, as a whole (emergent properties).[27]

Tools that aid in systems thinking include diagrams, flowcharts, maps, or pictures; these "work for this language better than words, because you can see all parts of a picture at once."[28] These graphical images are often referred to as "mind maps" or "situation maps." Here, they are referred to as *situation maps*; (Figure 3.2) provides an example. A situation map is a

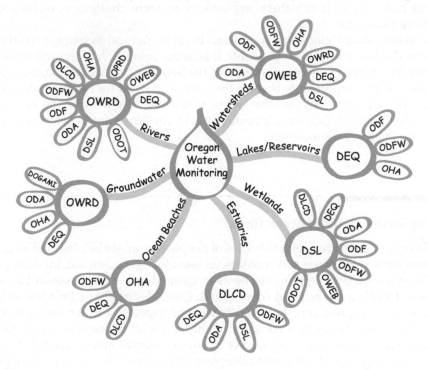

FIGURE 3.2

A situation map displaying the primary representation, leadership, and involvement of state agencies monitoring Oregon waters. (Drawing and interpretations by Janine Salwasser, Institute for Natural Resources, 2017.) OWRD is the Oregon Water Resources Department. OWEB stands for Oregon Watershed Enhancement Board. DEQ represents the Oregon Department of Environmental Quality. ODOT is the Oregon Department of Transportation, and DSL is the Department of State Lands. ODA is the Oregon Department of Agriculture. ODF denotes the Oregon Department of Forestry. ODFW is the Oregon Department of Fish and Wildlife. DLCD is the abbreviation for the Oregon Department of Land Conservation and Development. OHA is the Oregon Health Authority, OPRD represents the Oregon Parks and Recreation Department, and DOGAMI stands for the Oregon Department of Geology and Mineral Industries.

preliminary means of illustrating linkages among components (animals in the Serengeti or actors/stakeholders in a particular situation or central interest). It is a means of showcasing relationships, in addition to conflicts and cooperation, when relevant. It can illustrate what is coming into the system (input) and what is leaving (output). It can indicate what constitutes the system and its boundaries and how the feedback loops work. And, in cases involving environmental stakeholders, a situation map can give a glimpse of who is involved in the process and who is missing. It is also a tool that can showcase types of interaction (significant or less significant to the whole system) and can be utilized to identify weak links in system relationships. Consequently, situation maps are increasingly used to better understand complex community-based water issues. In addition, situation maps can be helpful in understanding family dilemmas, as well as natural systems, ecosystem challenges, or human-altered systems.

System observations are significant because natural systems have their own rhythm and pulse. And, note that some species, like the wildebeest, have far-reaching impacts throughout the Serengeti compared to other species.[29] So, it is important to observe the system and how the components interact and not focus only on what can be measured.[30] Remember, the *system is offering its knowledge*; our responsibility and challenge are to listen and correctly interpret the message.[31]

Exploring Group Interactions

Groups, as used here, refer to bodies of people that might be classified as government sectors, water stakeholders, an assemblage, or a crowd, for example. These associations tend not to be as integrated as a classic system but are more loosely arranged and coordinated. Examples might include the 30 US agencies and programs that oversee various aspects of water management in the United States or a consortium charged with overseeing a particular task. Such groups are a key aspect of environmental management because the meetings that drive action take place at this level. To elucidate how this works, we next discuss stakeholder engagement followed by international involvements.

Stakeholder Engagement

The 1964 Columbia River Treaty is regarded as an exceptionally effective agreement between the United States and Canada, addressing flood control and power benefits on the fourth-largest river in North America.[32] However, today there are many more basin concerns, such as population shifts to the

region; deteriorating infrastructure; climate change issues; greater demands
for power and energy; and the need to consciously, purposefully conserve
ecosystem sustainability.

In light of the possibility, as stipulated in the treaty, that the current
arrangement could be renegotiated after 2014, a nonpartisan regional group
of institutions, all of whom are in the basin, in 2008 formed the *Universities
Consortium of Columbia River Governance.*[33] The institutions comprise the
University of Idaho, Oregon State University, University of Montana,
Washington State University, and University of British Columbia. This
consortium was designed both to facilitate negotiation among basin stakeholders and to act as a mechanism to inform policy. To this end, between
2009 and 2012, the consortium held the forums the *Annual Symposium on
Transboundary River Governance in the Face of Uncertainty: The Columbia River
Treaty, 2014,* in Idaho, Oregon, and Montana states and in British Columbia,
Canada.[34] The fact that British Columbia hosted the symposium was significant in bringing together representatives from the United States and a
larger number from the Canadian part of the basin. This consortium gave
the US contingent another physical and metaphorical viewpoint of the
basin in addition to experiencing firsthand some of the values and interests
of northern communities.

Field trips can provide experiential and contemplative learning that can be
transformative and impactful. Such outings offer real-world exposure, context, and the opportunity to experience the natural setting being discussed.
Through these means, information becomes sensory, assisting clarity and
understanding that can take learning deeper.[35]

I (Lynette), along with other participants, took part in a symposium boat
excursion on the Koocanusa Reservoir (in British Columbia) that incorporated those aspects. This reservoir (constructed in association with the
Montana Libby Dam in the 1970s) was one of several structures commis-
sioned by the Columbia River Treaty to provide flood control and power to
the basin. While there were many beneficial aspects to this construction,
the locals experienced some of the negative effects. As such, the boat ride
provided a perspective of the impacts this particular development has had
on residents over the decades, which includes the variation of reservoir
levels affecting recreation; the livelihood of boatmen; the harmful effects
to the fish population and ecosystem; disruption of the local economy and
social aspects of the area; and the increasing dust storms when the water
levels in the reservoir are extremely low and the banks of the river are
exposed.

We also attended a public meeting held by the Columbia Basin Trust at the
nearby town of Jaffray, British Columbia, where we experienced more local
sentiment associated with the reservoir's construction and the outstanding
issues regarding it. When the meeting started, it clearly seemed like two distinct groups, those south of the Canada–US border and those north of it. Over
dinner and through small-group discussions, there emerged opportunities

for all to tap into our common humanity, giving way to talks of family, community, and concerns about the Columbia River.

It was a learning experience for all present to see the perceptions of each side and learn about the interests underlying the position of the Jaffray community. As we listened to our neighbors' point of view, conceptually it was as though the national, provincial, and state boundaries were removed. And, it was on a bulletin board at this meeting that the local community started to map a shared vision of their future and their hopes for water resource management. The notion of changing perceptions is not an easy one by any means. We participated in the beginning of a process where parties meet and listen to the needs of one another. Clearly, this process cannot be a one-time interaction, but a series of evolving conversations.

International Involvements

Organizational representatives come to Oregon State University from all over the world on matters of both technical and environmental management. A few years ago, Aaron Wolf and I (Lynette) had the distinct pleasure of hosting technical advisory committee members from the African Ministers' Council on Water. They held such positions as the technical advisor to the Ministry of Energy and Water; assistant director of transboundary waters for the Ministry of Water and Irrigation; and head of the water and sanitation department for the Ministry of Energy and Water. Even with their extensive work-related experience, which collectively spanned more than a century, they traveled like water diplomats from many nations to share and exchange expertise. These representatives were a specialized, technical committee of the African Union from several nations,[36] here to participate in a field trip and crash course in water conflict management.

We toured the Columbia River Gorge and the Bonneville Dam. We observed the panoramic and picturesque view of the Columbia River Gorge at Crown Point, while also getting a historical perspective of the area. It was here that one of the African delegates asked where the animals were (buffalo, rhinoceros, and perhaps zebra and other wildlife) that are typically found along his country's watercourse. Clearly, there are contrasts between the Columbia and the Nile Rivers. It was evident that we faced similar—and yet very different—water- and ecosystem-related challenges. There were opportunities to share information about the Columbia River system and the river treaty and showcase that Bonneville-Dam specialists comprised engineers, biologists, and economists, with public outreach and transparency a vital and important part of the program.

Their time at Oregon State University began with introductions: a standard introduction followed by either telling stories about the origin of their names, discussing an article of clothing they were wearing that was important to the individual, or telling their water story. It was interesting that, right when I thought we would move on to more *important* water matters, one of the delegates insisted on drawing us back to talk about the origin of his name and then asked Dr. Wolf to tell his story and me to tell mine. So, interactions have a dynamic of their own and cannot be rushed. And, it is important to give ample time to each stage in the process. It reaffirmed that, once people feel listened to, they would also be more open to listening.

We spent time getting to know what the individuals from the African Ministers' Council on Water hoped to learn while with us. It was important to first take the time to discover how we could help and then what they wanted to know. Although there was communication prior to their arrival through the coordinators of the African Ministers' Council on Water, it was good to allow the actual people to say what they wanted to get out of the forum. Sometimes, you find that it is actually not what you thought or had been led to think. It honors their presence. It is a great opportunity for instructors to practice their listening skills and to keep in mind that the learning process works both ways.

We learned that they were interested in focusing on improving communications at the international level in the transboundary process (improving interaction between countries in the same basin and region) and less at the intranational level. However, even with this knowledge, the workshop began with communication at the personal level because that is where all interactions lie. Think about it: Even when you are working at the international level, you are always interacting with a person representing an institution, agency, region, or nation.

There were some noticeable breakthrough moments that might be called shifts of awareness that took place, and there were opportunities to role-play as major stakeholders in an imaginary basin. Strategically, the role-playing was done so the African delegates in upstream parts of the watershed could experience what it was like to be a downstream stakeholder and vice versa. This role-playing laid the groundwork for mock negotiations from a perspective different from their professional daily viewpoint and from a perspective that could benefit the Nile basin and region as a whole. The skill set gained through this exercise was invaluable. When the workshop was over, I was privileged to sit in on their "real" meetings the next day and see how they used the tools they had just acquired to invigorate their own basin and African Ministers' Council on Water processes. It was very interesting.

Exploring Interpersonal Interactions

I (Lynette) started to see a pattern that could not be denied. On more than one occasion, I found myself at a gathering (water conference, social event) having a conversation with an individual. While conversing, I would notice another person with whom I desperately wanted to talk; this urge took my attention away from the individual with whom I was *engaged*. Bringing the conversation to a close, I would connect with the other person, only to discover that the individual I left was the one who could help me with the specific information I needed. Observing that this scenario was happening over and over, it became clear that my communication skills were not all they could be. I was missing the gifts of the moment. What was required of me was to be fully present and engaged in each exchange.

Staying connected in time and space is the essence of being present. One crucial aspect of this can be accomplished through our senses (touch, smell, hearing, sight, or taste). Jill Bolte Taylor suggested using the "normal anatomy of what you are as a living being in order to stimulate you and bring you back to the present moment."[37] For example, I (Lynette) consciously tighten muscles (the palms of my hand; touch),[38] use aroma (perfume; smell),[39] or select colorful clothing to wear (sight) to refocus me to "now." This is my approach to staying grounded to my dimensional positioning in time and space. This allows for recognizing a moment as unique and makes interactions (human, terrestrial, and spiritual) more conscious and alive.

Pema Chödrön said, "If we knew that tonight we were going to go blind, we would take a long, last real look at every blade of grass, every cloud formation, every speck of dust, every rainbow, raindrop—everything."[40] There is a tendency on the timescale to see this moment as infinitesimal and the before and after (of this moment) as expansive and linear. Chris Prentiss[41] and Eckhart Tolle[42] suggested that with *this instant* forever present, the moment to moment might be the expanse, so in reality, *this* is all there is.

During a seemingly mundane task, such as walking through the grocery aisle of a store, one might not expect to have a profound experience, but in 2017, I (Lynette) did, right there in the supermarket. It might have appeared to be a simple exchange. My eyes locked with the eye of a total stranger. The look was momentary and also lifelong, complete. No wink, no smile, no words were exchanged. The person remains to this day unidentifiable, as though blended back into everyday life, but I learned through that experience that we are more than the sum of our parts. One *supreme* glance of love between strangers is enough to carry us through time. I learned that all encounters are sacred. All meetings count.

There are many forms of communication (and many styles of conflict management) in which there might be tendencies for an individual to collaborate, negotiate a compromise, compete, avoid, or accommodate the perspectives of others.[43] No one necessarily adheres to the same approach for all interactions. However, a vital part of being in communion with another human being is to practice the art of listening.[44]

Listening is seldom fully valued (except by those wanting to be heard), so the idea of a class exercise focused on listening seldom gets students excited at first. Yet, after this kind of class activity, participants want more. The exercise requires a quiet place and the time commitment to be fully engaged with no distractions. The listener is asked to face the speaker and be fully present, at ease, making eye contact and, as the expression goes, *smile into one's heart*. The listener does not interrupt but provides support to the speaker through affirmative (silent) gestures that show interest and acknowledge that one is listening. If clarification is needed, encourage the speaker to elaborate on a particular aspect. And, toward the end, when the speaker has completed their thoughts, the listener can reflect on what was heard, without judgment, permitting the speaker to correct the listener if needed. This approach honors the speaker and gives unforeseen gifts to the listener (openness and intimacy).

Exploring Intrapersonal Interactions

In December 1996, Jill Bolte Taylor, a neuroscientist, sustained a brain hemorrhage. However, she was conscious enough to witness it while experiencing it. Though debilitating, she not only made a full recovery after 8 years but also documented her experiences and brain activity, thereby gaining a unique perspective of the functions of the left and right hemispheres of the brain. The loss of the left side of her brain left her unable to tap into analytical reasoning, mathematical functioning, verbal cues, linear thinking, and that sense of the individual self ("I"). Even so, she was able to fully experience the right hemisphere. This hemisphere, associated with holistic thinking, intuition, and artistic expression, allowed her to experience the universe in its oneness, as an integrated system, expressing inclusion and the collective, compassionate sense of "us." Through her trauma and recovery, she became more cognizant of the gifts that the right hemisphere provides and the necessity to consciously activate those aspects of the brain in order to stay rooted in union.[45]

After students watched Jill Bolte Taylor's TED talk,[46] I (Lynette) ventured to ask the class what could be gleaned from this experience. And, what did this have to do with environmental conflict management? Which of the hemispheres do you choose and when? Do we in fact choose? And, without having a stroke or taking drugs, what else could give one the understanding associated with the right hemisphere (that Jill had) to experience the essence of her discovery? I emphasized to the class that it is for each one of us to find our own answers to these questions. These are primarily rhetorical questions, and while some may still have been processing the information, I suggested that they only share their thoughts if they wanted to do so.

For me, the essence of Jill's experience includes being more connected to the universe and at peace; being temporarily disconnected to the endless whims of the mind; feeling less burdened by unprocessed feelings associated with past interactions and place; and feeling lighter in my body.

I think we can all relate to having had moments when our experiences (though perhaps not necessarily as powerful) come closer to Jill's experience. For me, it happens when I am meditating; being still; utilizing some of the same techniques I use before I take an examination to center myself; being present; taking slow, deep breaths; watching a sunset; losing myself in an artistic/creative pursuit; and being close to the ocean or to Nature.

Observe for yourself which techniques work for you and whether your day is different when your own techniques are used or *not* used. You can gauge for yourself what rippling effects this might have at home, work, and play and how far that rippling effect might extend. Consider the quality of the exchange when individuals are fully present, centered, and engaged. The significance of these techniques is to hear our calling, our duty, our charge, and the ability to be in communion so we can listen for what is needed (for our individual self, each other, the community, this space and moment in time and our planet).

This chapter demonstrates that meaningful relationships (complex, group, interpersonal, and intrapersonal) can be forged and transformed at any scale through the use and practice of tools. Instruments like using situation mapping; utilizing neutral and nonpartisan brokers and forums; taking field trips to inform communication; practicing the art of listening; and being fully present can be a means to strengthen cooperation and mitigate conflicts.

Discussion Questions

1. Think about a complex system with which you are familiar. What evidence is there that additional meaningful communication might be needed?

2. Think about group interactions with which you are familiar. What evidence is there that additional meaningful communication might be needed?

3. In what ways might you engage at the interpersonal scale to create more meaningful communication?

4. Make a list of five simple ways you could utilize your senses (touch, smell, hearing, sight, or taste) as a means to keep you connected in time and space. Over the course of a week, try them out. Of the approaches you listed, which ones proved most effective?

5. Try practicing the art of listening as an exercise. You can use the technique described in this chapter. How effective was the technique for you?

6. Is there a particular question you would like to ask?

Endnotes

1. Howard Wiseman. Physics: Bell's Theorem Still Reverberates. *Nature,* 510(2014):467–469.
2. (1) Jane Silberstein and Chris Maser. *Land-Use Planning for Sustainable Development.* 2nd ed. CRC Press, Boca Raton, FL, 2014. 296 pp.; (2) Sean B. Carroll. *The Serengeti Rules: The Quest to Discover How Life Works and Why It Matters.* Princeton University Press, Princeton, NJ, 2016. 263 pp.; (3) Deepak Chopra. *The Seven Spiritual Laws of Success: A Practical Guide to the Fulfillment of Your Dreams.* Amber-Allen, San Rafael, CA, 1994. 112 pp.
3. Chopra, *Seven Spiritual Laws.*
4. Carroll, *Serengeti Rules.*
5. Mathis Wackernagel, Niels B. Schulz, Diana Deumling, and others. Tracking the ecological overshoot of the human economy. *Proceedings of the National Academy of Sciences of the United States of America,* 99(2002):9266–9271.
6. (1) United Nations, Department of Economic and Social Affairs, Population Division. World Population Prospects: The 2017 Revision. https://www.compassion.com/multimedia/world-population-prospects.pdf (accessed September 20, 2017); (2) World Population Prospects 2017. https://esa.un.org/unpd/wpp/DataQuery/ (accessed March 1, 2018).
7. Wackernagel et al., Tracking the Ecological Overshoot.
8. Carroll, *Serengeti Rules.*
9. Hans Rosling. Why the world population won't exceed 11 billion. https://www.youtube.com/watch?v=2LyzBoHo5EI (accessed March 1, 2018).
10. United Nations, Department of Economic and Social Affairs, Population Division. World Population Prospects: The 2017 Revision. Custom data acquired via website. 2017. https://esa.un.org/unpd/wpp/DataQuery/ (accessed March 1, 2018).
11. Hans Rosling. Global Population Growth, Box by Box. TED (= Technology, Entertainment and Design converged) [Lecture]. June 2010. https://www.youtube.com/watch?v=fTznEIZRkLg (accessed March 1, 2018).
12. Carroll, *Serengeti Rules,* p. 10.
13. Peter Brown. Are There Any Natural Resources? *Politics and the Life Sciences,* 23(2004):12–21.
14. John C. Z. Woinarski and Alaric Fisher. The Australian Endangered Species Protection Act. *Conservation Biology,*13(1999):959–962.
15. Endangered Species Act of 1973. https://en.wikipedia.org/wiki/Endangered_Species_Act_of_1973 (accessed April 18, 2018).
16. Steven G. Davison. The Aftermath of Sweet Home Chapter: Modification of Wildlife Habitat as a Prohibited Taking in Violation of the Endangered Species Act, 27 Wm. & Mary Envtl. L. & Pol'y Rev. 541 (2003). http://scholarship.law.wm.edu/wmelpr/vol27/iss3/2 (accessed March 1, 2018).
17. Marvin W. Meyer. Introduction. In: Marvin W. Meyer and Kurt Bergel, Eds., *Reverence for Life: The Ethics of Albert Schweitzer for the Twenty-First Century.* (Syracuse University Press, Syracuse, NY, 2002, pp. xi–xvii. 350 pp.
18. Thomas A. Carr, Heather L. Pedersen, and Sunder Ramaswamy. Rain Forest Entrepreneurs: Cashing in on Conservation. *Environment, Science and Policy for Sustainable Development,* 35(1993):12–38.

19. Carroll, Serengeti Rules.
20. (1) Silberstein and Maser, *Land-Use Planning*; (2) Carroll, *Serengeti Rules*.
21. (1) Chris Maser. *Decision Making for a Sustainable Environment: A Systemic Approach*. CRC Press, Boca Raton, FL, 2013. 304 pp.; (2) Cameron La Follette and Chris Maser. *Sustainability and the Rights of Nature: An Introduction*. CRC Press, Boca Raton, FL, 2017. 418 pp.
22. (1) Maser, *Decision Making*; (2) Carroll, *Serengeti Rules*; (3) Michael Goodman. Systems Thinking: What, Why, When, Where, and How? https://thesystems-thinker.com/systems-thinking-what-why-when-where-and-how/ (accessed March 1, 2018).
23. Goodman, Systems Thinking.
24. Steven E. Daniels, and Walker B. Gregg. *Working Through Environmental Conflict: The Collaborative Learning Approach*. Prager (an imprint of Greenwood Publishing), Westport, CT, 2001. 328 pp.
25. Ibid.
26. Ibid.
27. Ibid.
28. Donella H. Meadows. *Thinking in Systems: A Primer*. Chelsea Green, White River Junction, VT, 2008. 240 pp.
29. Carroll, *Serengeti Rules*.
30. Meadows, *Thinking in Systems*.
31. Ibid.
32. Columbia River Treaty. https://en.wikipedia.org/wiki/Columbia_River_Treaty (accessed April 22, 2018).
33. Universities Consortium on Columbia River Governance. The Columbia River: A Sense of the Future. 2013. https://www.crt2014-2024review.gov/Files/Universities%20Consortium.pdf (accessed April 23, 2018).
34. Barbara Cosens. Transboundary River Governance in the Face of Uncertainty: Resilience Theory and the Columbia River Treaty. *Journal of Land Use & Environmental Law*, 30(2011). https://www.researchgate.net/publication/228185232_Transboundary_River_Governance_in_the_Face_of_Uncertainty_Resilience_Theory_and_the_Columbia_River_Treaty (accessed April 23, 2018).
35. Daniel Barbezat and Mirabai Bush. *Contemplative Practices in Higher Education: Powerful Methods to Transform Teaching and Learning*. Jossey-Bass, San Francisco, 2014. 231 pp.
36. African Ministers' Council on Water. Welcome to AMCOW Where Every Drop Counts. http://amcow-online.org/index.php?option=com_content&view=article&id=69%3Aabout-amcow&catid=34%3Aabout-amcow&Itemid=27&lang=en (accessed March 1, 2018).
37. Jill Bolte Taylor. Jill Bolte Taylor with Oprah 12 of 12 [Interview]. 2012. https://video.search.yahoo.com/search/video?fr=tightropetb&p=bolte+tayor+ophrah#id=2&vid=36c92b206e07c264878b013a3164de69&action=click (accessed March 1, 2018). [Video section 5:10–5:17.]
38. Ibid.
39. Ibid.
40. Pema Chödrön. *Offerings: Buddhist Wisdom for Every Day* (edited by Danielle Föllmi and Olivier Föllmi). Shambhala, Boulder, CO, 2003, p. 164.

41. Chris Prentiss. *Zen and the Art of Happiness*. Power Press, Malibu, CA, 2006. 145 pp.
42. Eckhart Tolle. *The Power of Now: A Guide to Spiritual Enlightenment*. Namaste, Novato, CA, 2004. 229 pp.
43. Aaron T. Wolf (Ed.). *Sharing Water, Sharing Benefits: Working Towards Effective Transboundary Water Resources Management*. United Nations Educational, Scientific and Cultural Organization, Paris, 2010. 278 pp.
44. Tolle, *Power of Now*.
45. Jill Bolte Taylor. 2008. "My Stroke of Insight". TED (= Technology, Entertainment and Design converged) [Lecture]. February 2008. https://www.youtube.com/watch?v=PzT_SBl31-s (accessed March 1, 2018).
46. Ibid.

4

The Global Commons and the Social Principles of Sustainability

Simply stated, a *commons* is something owned by everyone and so by no one. Moreover, the global commons are the "birthright" of every living thing—not just humans. From a human perspective, however, it is the vast realm of our shared heritage, which we typically enjoy and use free of toll or price. Air, water, and soil; sunlight and warmth; wind and stars; mountains and oceans; languages and cultures; knowledge and wisdom; peace and quiet; sharing and community; and the genetic building blocks of life—these are all aspects of the commons.

The commons have an intrinsic quality of just being there, without formal rules of conduct. People are free to breathe the air, drink the water, and share life's experiences without a contract, without paying a royalty, without needing to ask permission. The commons are simply waiting to be discovered and used.

For example, as a youth in the 1950s, and even as a young man in the earliest years of the 1960s, I (Chris) could stand on the shoulder of a mountain in the High Cascades of Oregon or Washington and gaze on a land clothed in ancient forest as far as I could see into the blue haze of the distance in any direction. My sojourns along the trails of deer and elk were accompanied by the wind as it sang in the trees and by the joy-filled sound of water bouncing along rocky channels. At other times, the water gave voice to its deafening roar as it suddenly poured itself into space from dizzying heights, only to gather itself once again at the bottom of the precipice and continue its appointed journey to the mother of all waters, the ocean.

Throughout those many springs and summers, the songs of wind and water were punctuated with the melodies of forest birds. Wilson's warblers sang in the tops of ancient firs, while the plaintive trill of the varied thrush drifted down the mountainside, and the liquid notes of winter wrens came ever so gently from among the fallen monarchs, as they lay decomposing through the centuries on the forest floor. From somewhere high above the canopy of trees came the scream of a golden eagle, and from deep within the forest emanated the rapid, staccato drumming of a pileated woodpecker. These were the sounds of my youth. This was the music that complemented the forest's abiding silence—a silence that archived the history of centuries and millennia as the forest grew and changed, like an unfinished mural painted with the infinite novelty of perpetual creation.

And so it is that every aspect of the commons engages people in the wholeness of themselves. It fosters the most genuine of human emotions and stimulates interpersonal relationships in order to share the experience, which enhances its enjoyment and archives its memory.

Human Population: A Matter of Gender Equality

We have been warned for decades that the human species is overpopulating Earth. Yet, our population explodes, and the usable portion of Earth per individual shrinks, as does the allotted proportion of its resources, all of which become more quickly limiting when abused. We have tried many things to remedy this situation: education, birth control, feeding the hungry, shipping industrial technology to poor nations, and so on. In our opinion, however, we have not addressed the primary cause of overpopulation: the staunchly maintained inequality between men and women.

Men have long dominated women. Through such domination, women are physically forced to produce most of the world's food yet are allowed to own but an infinitesimal part of the land. And, women have had only one way to be uniquely valued by men: by having babies, particularly *boys*.

Regardless of where I (Chris) have traveled, I have found that women who have a good education have fewer children and have them later in life. Education affords increased options and a variety of ways to be valued. If, therefore, we humans are serious about controlling our population, women must have an equal voice in all decisions and unequivocal access to opportunities for self- and social valuation. On the surface, this means such things as equal opportunity for education and jobs and equal pay for equal work. At its root, this means changing the male attitude of superiority toward women, a difficult task, but a vitally necessary one, because we humans directly affect such foundation stones of social survival: air, soil, water, biodiversity, and population density, as well as indirectly affecting sunlight.

If, for example, we choose to clean the world's air, we will automatically clean the soil and water to some extent because airborne pollutants will no longer poison them, and the sunlight that reaches Earth will be unimpeded. If we then choose to treat the soil in such a way that we can grow what we desire without the use of artificial chemicals, and if we stop using the soil as a dumping ground for toxic wastes and avoid overintensive agriculture, the soil can once again filter and purify water. If we stop dumping waste effluents into the water, streams, rivers, estuaries, and oceans could—with time—become cleaner and healthier.

With clean and healthy air, soil, and water, we can also have clear, safe sunlight with which to power Earth and have a more benign—and perhaps predictable—climate in which to live. In addition, a population in balance with its habitat will reduce demands on Earth's resources. With reduced competition for resources, cooperation and coordination can allow landscapes to provide maximum possible biodiversity. Protecting biodiversity translates into the gift of choice, which in turn translates into hope and dignity for future generations.

For the sake of discussion, let us add to this scenario the end of wars and their weapons. Such a world, a wonderful place in which to live and raise families, is possible, but this possibility ultimately hinges on clean air.

If we do everything outlined here except clean the air, we will still pollute the entire Earth, from the blue arc of its heavens to the bottom of its deepest sea, in every corner of the globe. Clean air is the absolute "bottom line" for human survival. Without clean air, there eventually will be no difference in the way we destroy ourselves, by either nuclear war or air pollution, because our world comprises three primary, interdependent, interactive spheres of our earthscape: the atmosphere (air), the lithohydrosphere (the rock that constitutes the restless continents and the water that surrounds them), and the biosphere (the life forms that exist within the other two spheres). We humans, however, arbitrarily delineate our seamless world into discrete ecosystems as we try to understand the fluid interactions between nonliving and living components of planet Earth. If you picture the interconnectivity of the three spheres as being analogous to the motion of a filled waterbed, you will see how patently impossible such divisions are because you cannot touch any part of a waterbed without affecting the whole of it. In other words, one affects the whole and the whole affects the one because everything is a relationship, and all relationships result in a transfer of energy (Biophysical Principles 1 and 5 in Chapter 5).

The paradox is that the thing of intangible value (such as scenic beauty) is the very thing that usually gives the most enjoyment to the greatest number of people over the longest time but turns no immediate profit. Commercial value, as opposed to intrinsic value, holds our mechanistic society captive; herein lies the conflict over the commons. Yet, we humans have jointly inherited the commons, which is more basic to our lives and well-being than either the market or the state, as observed by British economist and philosopher Edmund Burke:

> One of the first and most leading principles on which the commonwealth and the laws are consecrated is [that] the temporary possessors and life-renters in it [should be mindful] of what is due to their posterity . . . [and] should not think it among their rights to cut off the entail or commit waste on the inheritance by destroying at their pleasure the whole original fabric of society, hazarding to leave to those who come after them a ruin instead of a habitation.[1]

Burke's notion calls to mind a unifying construct of the commons, which can be thought of as the *Commons Usufruct Law*. *Usufruct* is a noun from ancient Roman law (and now a part of many civil law systems) that means one has the personal *right* to enjoy all the advantages derivable from the use of something that belongs to another, provided the substance of the thing being used is not injured in any way.[2] In Canada, for example, the indigenous First Nations people have a usufructuary right to hunt and fish without restriction on Crown lands. In a more industrialized setting, a farmer might rent an unused field to a neighbor, thus enabling that neighbor to sow and reap the harvest of that land or, perhaps, to use it as pasture for livestock.[3] On public rangelands in the western United States, the latter arrangement between the federal government and a local rancher is known as a *grazing allotment.*

The legal definition of *usufruct* in the United States is as follows:

> Usufruct is a right in a property owned by another, normally for a limited time or until death. It is the right to use the property, to enjoy the fruits and income of the property, to rent the property out and to collect the rents, all to the exclusion of the underlying owner. The usufructuary has the full right to use the property but cannot dispose of the property nor can it be destroyed.
>
> The extent of usufruct is defined by agreement and may be for a stated term, covering only certain stated properties; it could be set to terminate if certain conditions are met, such as marriage of a child or remarriage of a spouse; it can be granted to several people to share jointly; and it can be given to one person for a period of time and to another after some stated event occurs.[4]

How the Commons Usufruct Law Arose

Until AD 1500, hunter-gatherers occupied fully one-third of the world, including all of Australia, most of North America, and large tracts of land in South America, Africa, and northeast Asia, where they lived in small groups without the overarching disciplinary umbrella of a state or other centralized authority. They lived without standing armies or bureaucratic systems, and they exchanged goods and services without recourse to economic markets or taxation.[5]

With relatively simple technology (such as wood, bone, stone, fibers, and fire), they were able to meet their material needs with a modest expenditure of energy and have the time to enjoy what they had materially, socially, and spiritually. Although their material wants may have been few and finite and their technical skills relatively simple and unchanging, their technology was, on the whole, adequate to fulfill their requirements, a circumstance that says the hunting-gathering peoples were the original affluent societies.

The basic social unit of most hunting-gathering peoples, based on studies of contemporary hunter-gatherer societies, was the band, a small-scale nomadic group of 15 to 50 people who were related through kinship. These bands were relatively egalitarian in that leadership was rather informal and subject to the constraints of popular opinion. Leadership tended to be by example instead of arbitrary order or decree because a leader could persuade, but not command. This form of leadership allowed for a degree of freedom unknown in more hierarchical societies, but at the same time put hunter-gatherers at a distinct disadvantage when they finally encountered centrally organized colonial authorities.[6]

Hunter-gatherers were by nature and necessity nomadic—a traditional form of wandering, as a way of life, wherein people moved their encampment several times a year as they either searched for food or followed the known seasonal order of their food supply. "Home" *was* the journey in that belonging, dwelling, and livelihood were all components thereof. Home, in this sense, was "en route."

The nomadic way of life was essentially a response to prevailing circumstances, as opposed to a matter of conviction. Nevertheless, a nomadic journey is in many ways a more flexible and adaptive response to life than is living in a settlement.

This element of mobility was also an important component of their politics because they "voted with their feet" by moving away from an unpopular leader rather than submitting to that person's rule. Further, such mobility was a means of settling conflicts, something that proved increasingly difficult as people became more sedentary.

Nomads were, in many ways, more in harmony with the environment than a sedentary culture because the rigors and uncertainties of a wandering lifestyle controlled, in part, the size of the overall human population while allowing little technological development. In this sense, wandering groups of people tended to be small, versatile, and mobile.

Although a nomadic people may, in some cases, have altered a spring of water for their use, dug a well, or hid an ostrich egg filled with water for emergencies, they were largely controlled by when and where they found water. Put differently, water brought nomads to it. On the other hand, the human wastes were simply left to recycle into the environment as a reinvestment of biological capital each time the people moved on.

In addition, nomads, who carried their possessions with them as they moved about, introduced little technology of lasting consequence into the landscape, other than fire and the eventual extinction of some species of prey. Even though they may, in the short term, have depleted populations of local game animals or seasonal plants, they gave the land a chance to heal and replenish itself between seasons of use. Finally, the sense of place for a nomadic people was likely associated with a familiar circuit dictated by the whereabouts of seasonal foods and later pastures for their herds.

Another characteristic associated with mobility was the habit of hunter-gatherers to concentrate and disperse, which appears to represent the interplay of ecological necessity and social possibility. Rather than live in uniform size assemblages throughout the year, they tended to disperse into small groups, the aforementioned 15 to 50 people, that spent part of the year foraging, only to gather again into much larger aggregates of 100 to 200 people at other times of the year, where the supply of food, say an abundance of fish, made such a gathering possible.[7]

Although hunter-gatherers had the right of personal ownership, it applied only to mobile property, that which they could carry with them, such as their hunting knives or gathering baskets. Such fixed property as land, on the other hand, was to be shared equally through rights of use, but could not be personally controlled to the exclusion of others or the detriment of future generations.

Almost all hunter-gatherers, including nomadic herders and many village-based societies as well, shared a land tenure system based on the rights of common usage that, until recently, were far more common than regimes based on the rights of private property. In traditional systems of common property, the land is held in a kinship-based collective, while individuals owned movable property. Rules of reciprocal accesses made it possible for an individual to satisfy life's necessities by drawing on the resources of several territories, such as the shared rights among the indigenous Cherokee peoples of eastern North America.

In the traditional Cherokee economic system, both the land and its abundance would be shared among clans. One clan could gather, another could camp, and yet a third could hunt on the same land. There was a fluid right of common usage, rather than a rigid, individual right to private property. The value was thus placed on sharing and reciprocity, on the widest distribution of wealth, and on limiting the inequalities within the economic system.[8]

Sharing was the core value of social interaction among hunter-gatherers, with a strong emphasis on the importance of generalized reciprocity: the unconditional giving of something without any expectation of immediate return. The combination of generalized reciprocity and an absence of private ownership of land has led many anthropologists to consider the hunter-gatherer way of life as a "primitive communism," in the true sense of the word *communism*.

Hunter-gatherer peoples lived with few material possessions for hundreds of thousands of years and enjoyed lives that were, in many ways, richer, freer, and more fulfilling than ours. These peoples so structured their lives that they wanted little, needed little, and found what they required at their disposal in their immediate surroundings. They were comfortable precisely because they achieved a balance between necessity and want, by being satisfied with little. There are, after all, two ways to wealth: working harder or wanting less.

The !Kung bushmen of southern Africa, for example, spent only 12 to 19 hours a week getting food because their work was social and cooperative, which means they obtained their particular food items with the least possible expenditure of energy. Thus, they had abundant time for eating, drinking, playing, and general socializing. In addition, young people were not expected to work until well into their 20s, and no one was expected to work after age 40 or so.

Hunter-gatherers also had much personal freedom. There were, among the !Kung bushmen and the Hadza of Tanzania, either no leaders or only temporary leaders with severely limited authority. These societies had personal equality in that *everyone* belonged to the same social class *and* had gender equality. Their technologies and social systems, including their economies of having enough or a sense of "enoughness," allowed them to live sustainably for tens of thousands of years. One of the reasons they were sustainable is that they made no connection between what an individual produced and their economic security, so acquisition of things to ensure personal survival and material comfort was simply not an issue.[9]

The Precursor of Today's Environmental Conflicts

With the advent of herding, agriculture, and progressive settlement, however, humanity created the concept of "wilderness," so the distinctions between "tame" (meaning *controlled*) and "wild" (meaning *uncontrolled*) plants and animals began to emerge in the human psyche. Along with the notion of tame and wild plants and animals came the perceived need not only to "control" space but also to "own" it through boundaries in the form of landscape markers, pastures, fields, and villages. In this way, the uncontrolled land or wilderness of the hunter-gatherers came to be viewed in the minds of settled folk either as "free" for the taking or as a threat to their existence.

"One of the most important developments in the existence of human society was the successful shift from a subsistence economy based on foraging to one primarily based on food production derived from cultivated plants and domesticated animals."[10] Being able to grow one's own food was a substantial hedge against hunger and thus proved to be the impetus for settlement that, in turn, became the foundation of civilization. Farming gave rise to social planning, as once-nomadic tribes settled down and joined cooperative forces. Irrigation arose in response to the need for supporting growing populations—and so the discipline of agriculture was born. Around 5,000 BCE, the first cities were constructed in the southern part of the long valley near the Persian Gulf by an intelligent, resourceful, and energetic people who became known as the Sumerians. The Sumerians gradually extended their civilization northward over the decades to become the first great empire: Mesopotamia, the name given to this geographical area by the ancient Greeks, meaning "land between two rivers."[11]

Evidence indicates this early irrigation farming was accomplished through communally organized labor to construct and maintain the canals, which necessitated the scheduling of daily activities beyond individual households. Nevertheless, to support the inevitable increase in the local population required an economy wherein farming was combined with hunting and gathering. The commitment to agriculture was more than simply the transition to a sedentary life structured around sustainable, small-scale production of food; it was also the commitment to a set of decisions and responses that resulted in fundamental organizational changes in society; increased risks and uncertainties; and shifts in social roles as a result of the dependence on irrigation technology.[12]

As indicated by the necessity to schedule daily activities beyond individual households, agriculture brought with it both a sedentary way of life and a permanent change in the flow of living. Whereas the daily life of a hunter-gatherer was a seamless whole, a farmer's life became divided into *home* (rest) and *field* (work). While a hunter-gatherer had intrinsic value, as a human being with respect to the community, a farmer's sense of self-worth became extrinsic, both personally and with respect to the community, as symbolized by, and permanently attached to, "productivity," a measure based primarily on how hard they worked and thus the quantity of goods or services they produced. In addition, the sedentary life of a farmer changed the notion of "property."

So, the dawn of agriculture, which ultimately gave birth to civilizations, created another powerful, albeit unconscious, bias in the human psyche. For the first time, humans saw themselves as clearly distinct from and superior to the rest of Nature—in their reasoning at least. They therefore began to consider themselves as masters of, rather than members of, Nature's community of life. It seems to me (Chris) that farmers began to develop a mindset that was antibiodiversity with the birth of agriculture—an attitude that prevails among the world's farmers of today. In fact, wild Nature, humankind's millennial life support system, suddenly came to be seen as a fierce competitor, a perpetual enemy to be vanquished when possible and subjugated when not.[13]

Until fairly recently, historically speaking, property in Britannia, as early England was known, used to be a matter of possessing the right to use land and its resources, and most areas had some kind of shared rights. Today, the land itself is considered to be property, and the words for the British shared rights of old have all but disappeared: "estovers" (the *right* to collect firewood), "pannage" (the *right* to put one's pigs in the woods), "turbary" (the *right* to cut turf), and "pescary" (the commoner's *right* to catch fish) are no longer in the British vocabulary. Now, while the landowner's rights are almost absolute, the common people no longer have the right of access to most lands in England.[14]

Although a few cultures (such as Bedouin clans in the Middle Eastern deserts and the Lapland reindeer herders) still live lightly on the land, most of humanity leaves a heavy footprint, consuming nearly a quarter of Earth's biophysical productivity. In fact, land use continually transforms Earth's terrestrial surface, thereby resulting in changes within biogeochemical cycles and thus the ability of ecosystems to deliver services critical to human well-being.[15]

Thus, while the hunter-gatherers spontaneously created the Commons Usufruct Law in their living, it is today being progressively eroded by people, especially in the industrialized countries. To arrest this erosion, we must understand and accept that the quality of our individual lives depends on the collective outcome of our personal motives, decisions, and actions, as they coalesce in the environment over time, particularly with respect to our common inheritance.

It is today increasingly critical for us, in the technological society we are creating globally, to both understand and accept that the Commons Usufruct Law and its governing principles are subordinate to the biophysical principles of the *Law of Cosmic Unification*. And, just as the biophysical principles are inviolate in their governance of Nature, so the social principles of the Commons Usufruct Law are inviolate in their governance of a functional society wherein long-term social-environmental sustainability is the requisite outcome.

Social Principles of Engagement in a Sustainable Society

So, what at this juncture can we, as a society, *relearn* from the hunter-gatherers? We can relearn to live by the principles of social behavior embodied in the *Commons Usufruct Law*—principles that grew as unplanned, individual behaviors that simply emerged out of a life epitomized by the continual novelty of reciprocal relationships between humans and Nature over the millennia. We say *relearn* because, as writer Carlo Levi once said, "the future has an ancient heart."[16]

Social Principle 1: Sharing Life's Experiences Connects Us to One Another

We are compelled to share our life's experiences with one another, as best we can, to know we exist and have value, despite the fact that we are forever well and truly alone with each and every thought, each and every experience and the emotion it evokes in our personal journey from birth to death.[17] And, the best way to share experiences is by working together and taking care of one another along the way, which, incidentally, is the price of social-environmental sustainability.

Social Principle 2: Cooperation and Coordination Are the Bedrock of Sustaining the Social-Environmental Commons

When we couple cooperation and coordination with sharing and caring, it precludes the perceived need to compete for survival and social status, except in play—and perhaps storytelling. Linking individual well-being strictly to individual production is the road to competition—and ultimately conflict—which, in turn, leads inevitably to social inequality, poverty, and environmental degradation. Self-centeredness and acquisitiveness are not inherent traits of our species, but rather acquired traits based on a sense of fear and insecurity within our social setting, which foster the perceived need to impress others with our personal prowess: our ego's sense that more is always better, and *enough* does not exist.

The separation of work from the rest of life, which began centuries ago with the inception of agriculture, is today fully manifested. Jobs are so fundamental to people's sense of identity and class distinction within the social hierarchy that many become severely depressed when they lose their jobs. Moreover, today's competitive marketplace, where people are "bought" and "sold" at the economic convenience of businesses and corporations, is a breeding ground for stress-related illnesses due to the uncertainty and unpredictability of everyday life over which people feel increasingly out of control. Long-term stress not only wears down the body but also initiates the potential for high blood pressure, heart disease, mood disorders, and chronic pain brought on by relentless muscle tension.[18]

Social Principle 3: The Art of Living Lies in How We Practice Relationships

The art of living lies in how we practice relationships—beginning with ourselves—because that is all we ever do in life. Wisdom dictates, therefore, that we live leisurely, which means to afford the necessary amount of time to fully engage each thought we have, each decision we make, each task we perform, and each person with whom we converse in order to fulfill a relationship's total capacity for a quality experience (see Chapter 5's Biophysical Principles 1, 4, 5, and 11). We learn to live fully in the measure in which we learn to live leisurely, a sentiment echoed by Henry David Thoreau: "The really efficient laborer will be found not to crowd his day with work, but will saunter to his task surrounded by a wide halo of ease and leisure."[19] If one lives leisurely, all aspects of life blend into a seamless whole, wherein contentment and joy can be found. In this sense, everyone can be a lifelong artist of daily living if they so choose.

Social Principle 4: Success or Failure Lies in the Interpretation of an Event

On the one hand, all relationships are self-reinforcing feedback loops that are of neutral valuation in Nature because Nature has only intrinsic value. On the other hand, these same self-reinforcing feedback loops carry either a positive or a negative accent in human valuation because we want specific predetermined outcomes to give us the illusion of being in control of circumstances. This human dynamic is the same as that driving the notion of success or failure, wherein each is the interpretation of an event, but not the event itself.

I (Chris) was asked by a community in Northern California to mediate their conflict over how to "restore" their river to its previous condition. During the process, some of the longtime older residents began lamenting how "newcomers" into "their" valley had destroyed "their" river through years of overuse and abuse. Finally, a youth in his late teens, who had been in juvenile detention, spoke up and said, in effect, "I don't know what you're talking about. I've been working for 3 years with a crew to improve the river's condition. The river is so much better than when I started. What's your problem?"

Whereas the only thing the old-timers could see was the loss of what, to them, had been a "pristine condition," the boy perceived a vast improvement in a short period of time. The success or failure of the efforts to heal the river was perceived differently depending on the people's life experiences, as tailored by their perspectives. In this case, the old-timers constantly reinforced their collective grief over the *negative* changes for which they blamed others, whereas the young man was encouraged by his ability to nudge change in a *positive* direction, to which the old-timers were blind but for which he took credit.

Social Principle 5: You See What You Look for and Focus On

Is there more beauty and peace in the world than ugliness and cruelty? In the end, it may be a matter of what we choose to focus on: the beauty and kindness that surround us if only we look for it *or* the large spoonful of fear fed to us daily by the media. As Francis Bacon put it, "The best part of beauty is that which no picture can express."[20] Even in the midst of war, beauty exists in a smile, a hug, caring for a child, and the ever-fresh face of a flower. In essence, ordinary people are motivated primarily by their inner harmony and balance, which is expressed through their sense of aesthetics.

Social Principle 6: People Must Be Equally Informed If They Are to Function as a Truly Democratic Society

For a group of people to be socially functional, they must be equally and honestly informed about what is going on that affects them; in other words, there must be no secrets that are actually or potentially detrimental to any member. Inequality of any kind, based on gender or social class, is merely fear of inadequacy disguised as privilege.

Social Principle 7: We Must Consciously Limit Our "Wants"

By consciously limiting our "wants," we can have enough to comfortably fulfill our necessities, as well as some of our most ardent desires, and leave more for other people to do the same. In essence, there are two ways to wealth: want less or work more. Unfortunately, the capitalistic system of economics is based on dissatisfaction and a continual stimulus to purchase superfluous items at the risk of personal debt, the long-term expense of the environment, and thus growing impoverishment for all future generations.[21]

Social Principle 8: Every Decision Is the Author of a Never-Ending Story of Social-Environmental Outcomes for All Generations

With every act we consummate, we become the authors of a never-ending story, a mystery novel of everlasting change in the world. This is so because every thought put into action has an effect, and every effect is the author of yet another effect, ad infinitum (see Chapter 5's Biophysical Principles 8 and 9).

Social Principle 9: Simplicity Is the Key to Contentment, Adaptability, and Survival

Anyone can complicate life, but it requires genius to keep things simple. Simplicity in living depends on seeking things small, sublime, and sustainable. Moreover, simplicity is the key to contentment, adaptability, and survival as a culture; beyond some point, complexity becomes a decided disadvantage with respect to cultural longevity, just as it is to the evolutionary longevity of a species. As artist Hans Hoffman put it, "The ability to simplify means to eliminate the unnecessary so that the necessary may speak."[22]

Social Principle 10: Marvel at the Abundance and Resilience of Earth

The notion of scarcity is largely an economic construct to foster consumerism and increase profits but is not necessarily an inherent part of human nature. We need to overcome our fear of economically contrived scarcity and marvel instead at the incredible abundance and resilience of Earth and our sacred duty to protect it for all generations.

Social Principle 11: Only Mobile Property Can Be Owned Outright

Mobile property can be owned, whereas such fixed property as land, which can be borrowed long term, is to be shared equally through rights of generational use. In other words, a person can borrow land as a trustee but cannot personally own land to the detriment of any generation. After all, no human being on Earth can create land. Thus, no human has the right to degrade or destroy that which they cannot create and all coming generations must use.

Social Principle 12: Nature, Spirituality, and Human Well-Being Are Paramount

Placing material wealth, as symbolized by the money chase, above Nature, spirituality, and human well-being is the road to social impoverishment, environmental degradation, and the collapse of societies and their life support systems.

Social Principle 13: Every Legal Citizen Deserves the Right to Vote

Every legal citizen of every country deserves the right to an equal vote of their conscience on how their country is to be governed because they and their children and their children's children must live with the consequences of the collective choices and actions.

Social Principle 14: We Must Choose—In That We Have No Choice

As stated in Chapter 3, we have a choice in everything we think and almost everything we do, including practicing relationships, experiencing ourselves as we experience relationships, choosing, changing the world, living without killing, and dying. In those, we have *no* choice of what we do, but we *do* have a choice of how we do it, and we *must* choose because not to choose *is still a choice, a new choice*. In addition, we make a new choice (even if it is doing nothing) each time a circumstance in our life changes, which of course is an ongoing process in that each decision creates a kaleidoscope of additional choices. In turn, choice is the author of both wisdom and folly, which manifest as the consequences of our decisions and actions. This last statement is particularly relevant, as Israeli Statesman Abba Eban observed, "History teaches us that men and nations behave wisely once they have exhausted all other alternatives."[23]

Social Principle 15: We Change the World Simply Because We Exist

As an inseparable part of Nature, we have no choice but to change the world in our living simply because we exist and use the world's energy to survive (Chapter 5, Biophysical Principles 1 and 5). We do, however, have a choice in selecting the level of consciousness with which we treat our environment within the reciprocal relationships of life and living.

Social Principle 16: We Must Kill to Live

We must eat to live. To eat means we must kill or cause to be killed—whether plant or animal. If we choose not to eat in order not to kill, we kill ourselves through starvation. Therefore, we have no choice but to kill. Nevertheless, we do have a choice in the level of conscious awareness with which we decide what must be killed, why, and how to do so with the greatest humility and the least suffering we create in the process.

Social Principle 17: This Eternal, Present Moment Is All There Is

The present moment—the *eternal here and now*—is all there is. The past is a memory. The future never comes. We live in this eternal moment.

The foregoing social principles of engagement are enshrined in everyday life, whether we recognize them or not. Moreover, they form the basis of how Lynette and I practice conflict resolution through mediation and teaching.

It is therefore necessary to understand that change in behavior often occurs between feelings of need and fear. On the one hand, we know we need to do things differently; on the other hand, we are terrified of facing the unknown and unfamiliar. To change our direction in the present for the future, however, we must suspend our conventional notion about change and our ability to learn because there are no problems to resolve "out there." All problems and solutions lie within us: our social conditioning and consequential perspectives and perceptions that determine the consciousness with which we think and act. And, many people prefer to err again and again rather than let go of some cherished belief, pet notion, deified assumption, or entrenched point of view.

Discussion Questions

1. Why is the present moment, the here and now, important in resolving environmental conflicts?
2. Does the outcome of an environmental conflict in the present moment affect the "future"? If so, how?
3. Is there a particular question you would like to ask?
4. Provide an example of how Social Principle 8 might work?
5. Of the principles listed in this chapter, name one that resonates with you. Can you think of an example that seems most relevant?

Endnotes

1. Edmund Burke. http://www.conlaw.org/Intergenerational-II-2.htm (accessed on March 13, 2010).
2. *Webster's Unabridged Dictionary.* Random House, New York, 1999. 2230 pp.
3. Mike Fritz. Pastureland Survey Shows Lease Rates Still Climbing. *Beef Magazine,* February 23, 2007. http://beefmagazine.com/cowcalfweekly/pastureland-survey-lease-rates/ (accessed January 8, 2009).
4. USLegal. Usufruct Law and Legal Definition. http://definitions.uslegal.com/u/usufruct/ (accessed January 1, 2009).
5. The foregoing discussion is based on Richard B. Lee. Forward. In: John Gowdy, Ed., *Limited Wants, Unlimited Means.* Island Press, Washington, DC, 1998, pp ix–xii.
6. The foregoing discussion is based on (1) Lee, Forward;; (2) John Gowdy. Introduction. In: Gowdy, *Limited Wants,* pp. xv–xxix.
7. (1) Gowdy, Introduction (2) Marshall Sahlins. The Original Affluent Society. In: Gowdy, *Limited Wants,* pp. 5–41.
8. The foregoing two paragraphs are based on Rebecca Adamson. People who Are Indigenous to the Earth. *YES! A Journal of Positive Futures,* Winter (1997):26–27.
9. Gowdy, Introduction.
10. Tom D. Dillehay, Herbert H. Eling, Jr., and Jack Rossen. Preceramic Irrigation Canals in the Peruvian Andes. *Proceedings of the National Academy of Sciences of the United States of America,* 102(2005):17241–17244. Page 17241.
11. The preceding two paragraphs are based on (1) Stacey Y. Abrams. The Land between Two Rivers: The Astronomy of Ancient Mesopotamia. *The Electronic Journal of the Astronomical Society of the Atlantic,* 3(2) (no page numbers); (2) The Fertile Crescent. http://visav.phys.uvic.ca/~babul/AstroCourses/P303/mesopotamia.html (accessed January 7, 2009).
12. The preceding two paragraphs are based on Dillehay et al. Preceramic Irrigation Canals.
13. Wolfgang Haber. Energy, Food, and Land—The Ecological Traps of Humankind. *Environmental Science and Pollution Research,* 14(2007):359–365.
14. George Monbiot. Land Reform in Britain. *Resurgence,* 181(1997):4–8.
15. Helmut Haberl, K. Heinz Erb, Fridolin Krausmann, and others. Quantifying and Mapping the Human Appropriation of Net Primary Production in Earth's Terrestrial Ecosystems. *Proceedings of the National Academy of Sciences of the United States of America,* 104(2007):12942–12947.
16. Leonard W. Moss. Observations on "The Day of the Dead" in Catania, Sicily. *The Journal of American Folklore,* 76(1963):134–135. Page 134.
17. Chris Maser. *Of Paradoxes and Metaphors: Understanding Some of Life's Lessons.* Woven Strings, Amarillo, TX, 2003. 235 pp.
18. Radha Chitale. Job Loss Can Make You Sick. May 8, 2009. http://abcnews.go.com/Business/WellnessNews/story?id=7530730&page=1 (accessed June 1, 2009).
19. Odell Shepard. *The Heart of Thoreau's Journals.* Courier Dover, New York, 1961. 228 pp. (page 42).

20. Francis Bacon. http://Science.prodos.ORG (accessed January 2, 2009).
21. Russ Beaton and Chris Maser. *Economics and Ecology: United for a Sustainable World*. CRC Press, Boca Raton, FL, 2012. 191 pp.
22. Quotations Book. Quotes by Hans Hoffman. http://quotationsbook.com/author/3495/ (accessed January 7, 2009).
23. The Quotations Page. Abba Eban. http://www.quotationspage.com/quote/298.html (accessed January 10, 2009).

5

The Biophysical Principles of Sustainability

To resolve any dispute, the mediation process must go beyond human valuation of a resource to disclose and examine the fundamental issue of how use of the resource will affect the long-term social-environmental sustainability of the ecosystem of which it is a component. One must also recognize the long-term issues that need to be dealt with to resolve the long-term potential for destructive conflict. This is necessary because the environment and the sustainability of its resources are most often silent parties in disputes.

Among the exceptions to this are water bodies (such as New Zealand's Lake Waikaremoana and surrounding Te Urewera area and India's Ganga and Yamuna Rivers and their tributaries) that are being granted "legal identity," like those of people, with appointed representatives speaking on their behalf.[1] This measure provides more negotiating and legal stance and a wider arch of collective laws for building legal case if or when corrective action is needed. However, in most circumstances, Nature's sustainability, Earth's resources, and the next generations are silent parties in conflicts. Despite this, it proves difficult for current generations to act on and acknowledge an obligation to future generations.

To accommodate the sustainability of the environment's favorable ecological integrity—the "silent third party"—each person must understand Biophysical Principles 1–14 given further in this chapter as a condition for resolving a specific conflict. These are the inviolable, irreversible principles through which Nature operates and the philosophical valuations through which we must accept our participation with Nature. Understanding these principles of Nature's reciprocity means that whatever decisions are made in resolving a conflict are made consciously by those who can be held accountable for their outcome.

Social-environmental justice dictates that the participants of an environmental conflict be held accountable for the outcome of its resolution because the effects of their decisions become the consequences for all generations to come. This is particularly poignant in the face of an exploding human population, climate change, and rapidly dwindling resources.

The Paradox of Life

Herein lies a paradox: While we are compelled to share our life's experiences with one another, which constitutes a self-reinforcing feedback loop to know we exist and have value (see Biophysical Principle 2), we are forever well and truly alone with each and every thought, each and every experience, and the emotion they evoke in our personal journey from birth through death. A baby comes into the world from its mother's womb with its own experience of life within the womb and the birth process, something the baby can never share. An exception might be the birth of multiple siblings, such as twins, sharing a space-time continuum and DNA. On the other hand, the mother has her own experience of nurturing her child before birth, as well as the process of giving birth, which she can never share—not even with her child, albeit they coexisted for 9 months of their respective lives in the most intimate connection two human beings can have.

When we die, we pass out of life as we know it, but without the ability to share the experience, even when surrounded by family and friends. Therefore, we are born and we die *alone*—we are the only persons in the world who will ever truly experience the essence of who we are in our life as an individual.

Even if we could verbally share an experience with someone who had been through a similar situation, we would still be alone with our own rendition of it because all we can share are metaphors of our feelings and emotions through a chosen combination of symbolic words available in one's spoken language. In this sense, words can be likened to incandescent light bulbs because each bulb can have a different shape, clarity, color, hue, brightness of color, and intensity of the light based on wattage and can be energized by a 110- or 220-volt current, depending on the nation one is in (different shades of meaning based on the way individuals interpret such things as color and hue). Yet, all these various light bulbs share the commonality of electrical impulse: the energy (feelings and emotions) that fuel them. Moreover, our ability to share the exact meaning of the metaphors we choose depends on how conversant the person with whom we are visiting is with the language and the degree to which our experiential similarities coincide. We cannot, however, share the feelings, emotions, or thoughts themselves because they cannot be expressed directly through language, only through the metaphorical shadows it casts. Even two people in the midst of a deeply intimate sexual union have vastly different, private experiences, which neither can accurately portray to the other.

If the notion of being alone is expanded into the arena of life, it soon becomes apparent that we are alone with each thought we have, each question we ask, each decision we make, each rainbow or flower we observe, each bird's song we hear or symphony we listen to, and each emotion we feel. We are alone—totally alone—within a psychological world of our own making, regardless of how extroverted or introverted we are. Be it a world of exceeding beauty or terrific horror, we are the sole creator of the life we experience, and we live it alone, both as creator of our thoughts and as prisoner of our thinking.[2]

Our *aloneness*, in all its forms, is the essence of spiritual union with the Eternal Mystery, which transcends all words and their contrived meaning, because our solitude pares life down to only what is self-created and nothing more. Thus, solitude itself transcends all material understanding and in so doing is touched by the unifying paradox of the cosmos, namely, that all things are interrelated and alone—alone and interrelated.

In October 2017, my husband, Shan, and I (Lynette) were entertaining and enjoying the company of friends over dinner. The atmosphere was jovial, with friends telling amusing life experiences. Shan told a story that I knew well. Right after he finished telling the account, I thought to myself, "Wow, he really remembers a lot of the details; he recounted that story very well." It was a humorous tale, everyone laughed and so did I. In the midst of laughing, I thought about the original event and suddenly realized that he had not been there. In fact, I had told him the story, some time ago; yet, here he was artfully detailing the account, as I would have done it, as though he had been present. How is that possible? Chen et al.,[3] in their journal article, "Shared Memories Reveal Shared Structure in Neural Activity across Individuals," presented brain activity research (involving functional magnetic resonance imaging [fMRI]). They provided supporting evidence that relayed memories imparted to others mobilize the language components of the brain, sparking similar brain activity in listener and speaker. In so doing, the abstract becomes reality; shared memories become communal experiences.[4] And, we (literally) become witnesses for each other.

The *reality of life goes beyond words* because we cannot communicate all our feelings through language and because all that is real for us is how we feel. And, it is precisely because life's reality is silent that we need to care for something in which the language of the intellect is unnecessary, but the language of the heart, conveyed through touch, is vital, such as cohabiting with another person, raising a child, nurturing a pet, planting and tending a garden, engaging in caring for the commons, or resolving a conflict.

Astronomer and author Carl Sagan pointed correctly to cosmic unity when he said, "In order to make an apple pie from scratch, you must first create the universe,"[5] which includes such gifts of Nature as clean air, pure water, fertile soil, a rainbow, northern lights, a beautiful sunset, or a 5,000-year-old bristlecone pine growing in a national park. Nature's commons, said author Jonathan Rowe, is the "hidden economy, everywhere present but rarely noticed."[6] Nevertheless, Nature's biophysical services provide the basic social-environmental support systems of all life and well-being, such as air, soil, water, biodiversity, and sunlight, among others.

Air: The Breath of Life—And of Death

It is late afternoon on a clear, warm, sunny September day. A tiny spider climbs a tall stalk of grass in a subalpine meadow and raises its body into the air, almost standing on its head. From spinnerets on the tip of its abdomen, it ejects a mass of silken threads into the breeze. Suddenly, without visible warning, the spider is jerked off its stalk and borne skyward to join its relatives riding the warm afternoon air flowing up the mountainside, all casting their fortunes to the wind. Like their ancestors in centuries past, they float on air currents from the far corners of Earth, and some become the first inhabitants of newly formed South Seas islands.[7]

Spiders are not the only things borne aloft on air currents. On August 26–27, 1883, Krakatoa (a small Indonesian island between Java and Sumatra) was virtually obliterated by explosive eruptions that sent volcanic ash high enough above Earth to ride the world's airways for more than a year. This affected the climate by reducing the amount of sunlight reaching Earth, which in turn cooled the climate and affected all life.[8] Like the volcanic ash of Krakatoa, air also carries the reproductive spores of fungi and the pollen of various trees and grasses, as well as dust and microscopic organisms. And, it carries life-giving oxygen and water and death-dealing pollution—the legacy of human society. (For a thorough discussion of air as a system of transportation, see *Interactions of Land, Ocean and Humans: A Global Perspective.*[9])

Air can therefore be likened to the key in a Chinese proverb: To every man is given the key to the gates of heaven, and the same key opens the gates of hell. In this case, air is the key that carries both life-giving oxygen and death-dealing pollution, and as pollution increases, the quality and utility of the air decreases, which exemplifies the fact that everything has one or more trade-offs (see Biophysical Principle 7).

Soil: The Great Placenta

Soil is like an exchange membrane between the living (plant and animal) and nonliving components of the landscape. Derived from rock and organic matter, soil is built up by plants that live and die in it. It is also enriched by animals that feed on plants, void their bodily wastes, and eventually die, decay, and return to the soil as organic matter. Soil is by far the most alive and biologically diverse part of any terrestrial ecosystem. In addition, soil organisms are the regulators of most processes that translate into soil productivity.

Many cultures emphasize in their religion and philosophy that humans must be trustees of the soil. Confucius saw in Earth's thin mantle the sustenance of all life and the minerals treasured by human society. A century later, Aristotle viewed the soil as the central mixing pot of air, fire, and water that formed all things.

Most people cannot grasp these intangible, long-held beliefs. Thus, we pay little attention to the soil because it is as common as air and, like air, is taken for granted. But, if we pause and think about it, we see that human society is tied to the soil for reasons beyond measurable materialistic wealth.

Yet, in the name of short-term profits, people rob the soil of the very organic material necessary to its sustainable fertility. They also use artificial chemicals that poison the soil, alter the way its many hidden processes function, and pollute the water moving through it into the world's water system, including the ocean. Here, it must be understood that the sustainability of every system is defined by the integrity of its overall function (Biophysical Principle 4).

Soil is the stage on which the entire human drama is enacted. If we continue to destroy the stage on which we depend for life, we will play a progressively dominant role in a terminal tragedy of human society.

Water: A Captive of Gravity

Water and oxygen are the most important products produced from the world's forests. Most of our usable water comes from snows high on forested mountain slopes. When snow melts, the water percolates through the soil. It is purified when flowing through healthy soil; it is poisoned when flowing through soil stripped of Nature's processes and polluted with artificial chemicals. In addition, water bearing tons of toxic effluents flows directly into streams, rivers, estuaries, and the open ocean. Because water is a captive of gravity, all the pollutants it accumulates on its downhill journey eventually end up in the oceans, where they accumulate in ever-increasing concentrations,[10] illustrating that all relationships are novel and thus irreversible (see Biophysical Principles 8 and 9).

Biodiversity: The Variety of Life

Because every ecosystem adapts in some way to changes in its environment, biodiversity acts as an ecological insurance policy for the flexibility of future human options, which mirrors the fact that all systems are based on composition, structure, and function (see Biophysical Principle 10). In turn, the degree of a system's adaptability depends on the richness of its biodiversity, which creates backups: duplication or repetition of the elements of a system. Backups provide alternative functional channels in case of a failure and so retain the ability of a system to respond to continual change.

Each ecosystem contains backups (two to three species with similar biophysical functions) that provide resilience to absorb change or to bounce back after disturbance. Biological backups strengthen the ability of an ecosystem to retain its integrity. This means that the loss of one or two species is not likely to result in such severe functional disruption that it causes a total shift in how an ecosystem functions because other species can make up for the functional loss.

At some point, however, the loss of one or more species will tip the balance and cause an irreversible change that can lower the quality and productivity of the system with respect to the vital service the system provides us humans. This point of irreversibility is an unknown biological threshold, in that we do not know which species' extinction will disrupt or eliminate the services we require, which is why it pays to save every species possible. Hence, the following precautionary principle: Err on the side of biophysical prudence—*not* economic gain.[11]

In Dr. Seuss's children's book, *The Lorax*, a boy is charged with protecting and nurturing the growth of the last existing seed of the Truffula tree after loggers, industry, and humanity have pillaged the region, devastating the environment, and severely altering the ecosystem.[12] In real life, doomsday seed banks, such as the Svalbard Global Seed Vault, located about 808 miles (1,300 kilometers) from the North Pole in Norway, preserve Earth's seeds (Figure 5.1) so that, in the event of a world crisis, there will be food and some biodiversity.

Species variety is important because each species has a shape and structure that allow certain functions to take place. These functions interact with those of other species to create a viable system. All biodiversity is ultimately governed by the genetic code, which builds some backups into each ecosystem by replicating some degree of the character traits of a species.

Although a sustainable native ecosystem may respond positively to disturbances to which it is adapted, it may be vulnerable to the introduction of foreign disturbances to which it is not adapted. Indigenous plant and animal diversity therefore buffers an ecosystem against disturbances from which it cannot recover. With the local extirpation, or outright extinction, of species, we lose not only their contribution of structural and functional diversity but also their genetic diversity, which eventually results in complex ecosystems becoming simplified and unable to sustainably produce the services for which we valued them.

A forest, for example, is a living entity that often completes a cycle of interdependent processes over several centuries, spanning many human generations. Yet, with grossly incomplete, shortsighted knowledge and unquestioning faith in that knowledge, we predict the sustained-yield capability of economically designed tree plantations far into a problematic and unforeseeable future.

Based on these erroneous predictions, clear-cutting the old-growth forest and converting it to a biologically simplified plantation is traditionally justified economically, completely ignoring biodiversity, especially that which sustains the infrastructure of the forest soil. Destroy the soil and the forest ceases to be. Destroy the forest and the soil becomes further impoverished and erodes, which degrades water quality and diminishes the oxygen content of the air.

FIGURE 5.1
Entrance to the Global Seed Vault on Svalbard archipelago, halfway between Norway and the North Pole. (Photograph by Tiq.)

Sunlight: The Source of Global Energy

The sun has been worshipped for millennia, and the quality of its light has often been taken for granted. Sunlight, the only true investment of global energy, powers most of Earth's processes (Biophysical Principle 3); we harvest the sun's energy through the fruits and vegetables we eat. But, what happens when air pollution—smog—reduces the intensity of sunlight before it reaches plants? What happens when the ozone shield is compromised, allowing deadly ultraviolet rays to bombard Earth? What happens when both of these events occur simultaneously, as is now happening?

Impacts from our burgeoning human population, such as air pollution, now directly affect the quality of the sun's light and energy reaching us. Like everything else, the quality of the sun's light is a biophysical variable that must be taken into account.

The Waterbed Principle

As long as the human population was but a small fraction of its current size, Earth's resources were considered unlimited. But even then, "The history of almost every civilization," observed British historian Arnold Toynbee, "furnishes examples of geographical expansion coinciding with deterioration in [environmental and, therefore, resource] quality."[13] Think of this interconnectedness as the waterbed principle, which simply demonstrates that you cannot touch any part of a filled waterbed without affecting the whole of it.

Today, there is much talk about "renewable" resources, but no longer so much about unlimited resources. Ultimately, all physical resources are finite. Not only can we literally run out of a resource by exhausting its earthly supply, such as the extinction of a species and its attendant ecological function but also we can alter an existing resource to the point it is rendered useless to us, such as poisoning our drinking water through the myriad types and routes of pollution. And, we are doing both.

As the burgeoning human population demands more and more material commodities from a rapidly dwindling supply of raw materials, the ratio of resources apportioned to each human must decline. Further, those resources currently deemed renewable are only renewable as long as the system that produces them retains its functional integrity and is used solely in a biophysically sustainable manner.

Consider, for example, that as you hike in a wilderness area, wander through a national park, or take an oceanic cruise, no matter how far removed you seem to be from the center of society, you are still breathing pollution. It is everywhere and will continue to worsen as long as decisions to placate big industry continually trump a decisive, global pursuit of dramatically cleaning the world's air.

We dare not kid ourselves about the importance of air quality. Our earthly survival, and progressively that of our children and their children unto all generations, ultimately depends on clean air. Air is the interactive thread connecting soil, water, biodiversity, human population density, sunlight, and climate. (*Biodiversity* refers to the variety of living species and their biological functions and processes.) This interactive thread exemplifies the aforementioned waterbed principle.

Yet, we, as a society, with our myriad data bits and seemingly vast knowledge, listen to the world's economists and *assume* they are correct when they take such ecological variables as air, soil, water, sunlight, biodiversity, genetic diversity, climate, and more and convert them, in theory at least, into economic constants whose values are unchanging. Ecological variables are therefore omitted from consideration in our economic and planning models, and even from our thinking. Biodiversity and genetic diversity, on the other hand, are simply discounted when their consideration interferes with monetary profits; they are euphemistically termed *externalities*.[14] On top of it all is the nagging problem of human population growth. We talk about it, and worry about it, but in the end, we give only lip service to the one solution that can control it: total, real equality for women in all realms of life.

That notwithstanding, relationships among things are in constant flux as complex systems arise from subatomic and atomic particles in the giant process of evolution on Earth. In each higher level of complexity and organization, there is an increase in the size of the system and a corresponding decrease in the energies holding it together. Put differently, the forces that keep evolving systems intact, from a molecule to a human society, weaken as the size of the systems increases, yet the larger the system is, the more energy it requires in order to function. Such functional dynamics are characterized by their diversity as well as by the constraints of the overarching biophysical laws and subordinate principles that govern them.

These principles can be said to *govern* the world and our place in it because they form the behavioral constraints without which nothing could function in an orderly manner. In this sense, the Law of Cosmic Unification—the supreme law—is analogous to the Constitution of the United States, a central covenant that informs the subservient courts of each state about the acceptability of its governing laws. In turn, the Commons Usufruct Law (discussed in Chapter 4) represents the state's constitution, which instructs the citizens of what behavior is acceptable within the state. In this way, Nature's rules of engagement inform society of the latitude whereby it can interpret the biophysical principles and survive in a sustainable manner.

Understanding the Law of Cosmic Unification

The *Law of Cosmic Unification* is functionally derived from the synergistic effect of three universal laws: (1) the first law of thermodynamics, (2) the second law of thermodynamics, and (3) the law of maximum entropy production.

The *first law of thermodynamics* states that the total amount of energy in the universe is constant, although it can be transformed from one form to another. Therefore, the amount of energy remains entirely the same, even if you could go forward or backward in time. For this reason, the contemporary notion of either "energy production" or "energy consumption" is a non sequitur. The *second law of thermodynamics* states that the amount of energy in forms available to do useful work can only diminish over time. Thus, the loss of available energy to perform certain tasks represents a diminishing capacity to maintain order at a certain level of manifestation (say a tree) and so increases disorder or entropy. This "disorder" ultimately represents the continuum of change and novelty: the manifestation of a different, simpler configuration of order, such as the remaining ashes from the tree when it is burned. In turn, the *law of maximum entropy production* says that a system will select the path(s) out of available paths that maximizes entropy at the fastest rate given the existing constraints.[15]

The essence of maximum entropy simply means that, when any kind of constraint is removed, the flow of energy from a complex form to a simpler form speeds up to the maximum allowed by the relaxed constraint.[16] As it turns out, the law of maximum entropy production freed early hominids from one of the basic constraints of Nature when they adapted the intense entropy of burning wood to their everyday use. (A hominid, *hom·i·nid*, is any of the modern or extinct primates that belong to the taxonomic family Hominidae, *Hom·in·idae*, of which we are members.) Control of fire gave hominids the ability to live in habitats that heretofore had been too cold or where the seasonal temperature variations had been too great. It also allowed them to cook food, making parts of many plants and animals palatable and digestible when they were baked, roasted, or boiled. The charred remains of flint from prehistoric firesides on the shore of an ancient lake near the river Jordan in Israel indicate that our ancient ancestors had learned how to create fire 790,000 years ago.[17] Moreover, the increased supply of protein embodied in cooked meat is thought to have facilitated evolution of increasing hominid brain capacity, ultimately leading to our mental abilities.[18]

We, in contemporary society, are all familiar with the basis of this law even if we do not understand it. For example, we all know that our body loses heat in cold weather, but our sense of heat loss increases exponentially when windchill is factored into the equation because our clothing has ceased to be as effective a barrier to the cold—a constraint to the loss of heat—as it was before the wind became an issue. Moreover, the stronger and colder the wind, the faster our body loses its heat—the maximum entropy (distribution) of our body's energy whereby we stay warm. If the loss of body heat to windchill is not constrained, hypothermia and death ensue, along with the beginnings of bodily decomposition, reorganization from the complex structure and function toward a simpler structure and function. In other words, systems are by nature dissipative structures that release energy by various means but inevitably by the quickest means possible.

Now, let us examine the notion of energy dissipation in a more familiar way. I (Chris) have a wood-burning stove in my home with which I heat the 1,300 square feet of my living space. To keep my house at a certain temperature, I must control the amount of energy I extract from the wood I burn. I do this in nine ways.

Among my first considerations are the kind of wood I choose (such as Douglas fir, western red cedar, western hemlock); my choice is important because each kind of wood has a different density and thus burns with a corresponding intensity. Hardwoods require more oxygen to burn than the softwoods and burn longer with less intensity. A second concern is the quality of wood that I burn; well-seasoned wood burns far more efficiently than either wet, unseasoned wood or wood that is partially rotten. A third determination is the size and shape of the wood because small pieces produce a lot of heat but are quick to disappear, while large pieces take more time to begin burning, but last longer and may or may not burn as hot when they really get going, depending on the kind of wood. A fourth decision is how wide to open the damper and thereby control the amount of air fanning the flames, thus either increasing the intensity of burning (opening the damper) or decreasing the rate of burn (closing the damper). The wider the damper is opened, the less the constraint there is, and the hotter and faster the wood will burn and the more rapidly heat will escape: the *law of maximum entropy production*. This law also addresses the speed with which wood is disorganized as wood and reorganized as ashes.

The fifth choice is how warm I want my house to be in terms of how cold it is outside. A sixth consideration is how well my house is insulated against the intrusion of cold air and thus the escape of my indoor heat. A seventh option is how often I open the outside door to go in and out of my house and so let cold air flow. And, an eighth alternative is when to heat my house and for how long.

All eight variables are influenced by how warmly I choose to dress while indoors. Whatever I wear constitutes a constraint to heat loss of a greater or lesser degree. Clearly, the warmer I dress, the less wood I must burn in order to stay warm and vice versa. It is the same with how many blankets I have on my bed during the winter.

These nine seemingly independent courses of action coalesce into a synergistic suite of seemingly independent relationships, wherein a change in one automatically influences the other eight facets of the speed that energy from the burning wood escapes from my house. This said, the first and second laws of thermodynamics and the law of maximum entropy production meld to form the overall unifying law of the universe—the Law of Cosmic Unification—wherein all subordinate principles, both biophysical and social, are encompassed. With respect to the functional melding of these three laws, Rod Swenson of the Center for the Ecological Study of Perception and Action, Department of Psychology, University of Connecticut, said these three laws of thermodynamics "are special laws that sit above the other laws of physics as laws about laws or laws on which the other laws depend."[19]

Stated a little differently, these three laws of physics coalesce to form the *supreme* Law of Cosmic Unification, to which all biophysical and social principles governing Nature and human behavior are subordinate, yet simultaneously *inviolable*. Inviolable means that we manipulate the effects of a principle through our actions on Earth, but we do not—and cannot—alter the principle itself.

The Inviolable Biophysical Principles

Although I (Chris) have done my best to present the principles in a logical order, it is difficult to be definitive because each principle forms an ever-interactive strand in the multidimensional web of energy interchange that constitutes the universe and our world within it. Moreover, a different possible order can be found each time they are read, and each arrangement seems logical. Because each principle affects all principles (like a waterbed), every arrangement is equally correct:

Biophysical Principle 1: Everything Is a Relationship

The universe is a single relationship integrated by the eternal flow of energy in all its many facets. It is constituted of an ever-expanding web of biophysical feedback loops, each of which is forever novel and self-reinforcing. Moreover, each feedback loop is a conduit whereby energy is moved from one place, one dimension, and one scale to another. And, all we humans do—ever—is practice relationships within this web because the existence of everything in the universe is an expression of its relationship to everything else within the universal web. Moreover, all relationships are forever dynamic and thus constantly changing, novel, from the wear on your toothbrush from daily use to the rotting lettuce you forgot in your refrigerator. Herein lies one of the foremost paradoxes of life: The ongoing process of change is an irreversible, universal constant over which we have no control, much to our dismay.

Think, for example, what the difference is between a motion picture and a snapshot. Although a motion picture is composed of individual frames (instantaneous snapshots of the present moment), each frame is entrained in the continuum of time and thus cannot be held constant, as Roman Emperor Marcus Aurelius observed, "Time is a river of passing events, and strong is its current. No sooner is a thing brought to sight than it is swept by and another takes its place, and this too will be swept away."[20]

Yet we, in our fear of uncertainty, are continually trying to hold the circumstances of our life in the arena of constancy, as depicted in a snapshot, hence the frequently used term *preservation* in regard to this or that ecosystem, this or that building. Yet, jams and jellies are correctly referred to as "preserves" because they are heated during their preparation in order to kill all living organisms and thereby theoretically prevent noticeable change in their consistency.

Insects in amber are an example of the truest preservation in Nature. Amberization, the process whereby fresh resin is transformed into amber, is so gentle that it forms the most complete type of fossilization known for small, delicate, soft-bodied organisms, such as spiders and insects. In fact, a small piece of amber found along the southern coast of England in 2006 contained a 140-million-year-old spider web constructed in the same orb configuration as that of today's garden spiders. This is 30 million years older than a previous spider web found encased in Spanish amber. The web demonstrates that spiders have been ensnaring their prey since the time of the dinosaurs. And, because amber is three dimensional in form, it preserves color patterns and minute details of the organism's exoskeleton, allowing the study of microevolution, biogeography, mimicry, behavior, reconstruction of the environmental characteristics, the chronology of extinctions, paleo-symbiosis,[21] and molecular phylogeny.[22] But, the same dynamic cannot be employed outside an airtight container, such as a drop of amber or canning jar. In other words, whether natural or artificial, all functional systems are open because they all require the input of a sustainable supply of energy in order to function; conversely, a totally closed, functional system is a physical impossibility.

Biophysical Principle 2: All Relationships Are Productive

I (Chris) have often heard people say that a particular piece of land is "unproductive" and needs to be "brought under management." Here, it must be rendered clear that every relationship is productive of an outcome that has an *effect* in space and time, and the effect, which is the cause of another outcome, *is* the product. Therefore, the notion of an unproductive parcel of land or an unproductive political meeting is an illustration of the narrowness of human valuation because such judgment is viewed strictly within the extrinsic realm of personal values, usually economics, not the intrinsic realm of Nature's dynamics that not only transcend our human understanding but also defy the validity of our economic assessments.

We are not, after all, so powerful a natural force that we can destroy an ecosystem because it still obeys the biophysical principles that determine how it functions at any point in time. Nevertheless, we can so severely alter an ecosystem that it is incapable of providing—for all time—those goods and services we require for a sustainable life. To illustrate with a couple of examples, (1) the total surface area of the United States covered in paved roads precludes the soil's ability to capture and store water and (2) we are currently impairing the global ocean's ability to sequester carbon dioxide (one of the main greenhouse gases) because we have so dramatically disrupted

the population dynamics of the marine fishes by systematically overexploiting too many of the top predators.[23] All of the relationships that we affect are
productive of some kind of outcome—a product. Now, whether the product
is beneficial for our use or even amenable to our existence is another issue.

Biophysical Principle 3: The Only True Investment Is Energy from Sunlight

The only true investment in the global ecosystem is energy from solar
radiation (materialized sunlight); everything else is merely the recycling of
already existing energy. In a business sense, for example, one makes money
(economic capital) and then takes a percentage of those earnings and *recycles*
them, puts them back as a cost into the maintenance of buildings and equipment in order to continue making a profit by protecting the integrity of the
initial outlay of capital over time. In a business, one recycles economic capital
after the profits have been earned.

Biological capital, on the other hand, must be "recycled" *before* the profits
are earned. This means foregoing some potential monetary gain by leaving
enough of the ecosystem intact for it to function in a sustainable manner. In
a forest, for instance, one leaves some proportion of the merchantable trees
(both alive and dead) to rot and recycle into the soil and thereby replenish
the fabric of the living system. In rangelands, one leaves the forage plants in
a viable condition so they can seed and protect the soil from erosion, as well
as add organic material to the soil's long-term ecological integrity.

People speak incorrectly about fertilization as an *investment* in a forest
or grassland, when in fact it is merely recycling chemical compounds that
already exist on Earth. In reality, people are simply taking energy (in the
form of chemical compounds) from one place and putting them in another
for a specific purpose. The so-called investments in the stock market are a
similar shuffling of energy.

When people *invest* money in the stock market, they are really recycling
energy from Nature's products and services that were acquired through
human labor. The value of the labor is transferred symbolically to a dollar
amount, thereby representing a predetermined amount of labor. Let us say
you work for $10 an hour; then, a $100 bill would equal 10 hours of labor.
Where is the *investment*? There is not investment, but there is a symbolic
recycling of the energy put forth by the denomination of money we spend.

Here, you might argue that people *invested* their labor in earning the
money. And, I (Chris) would counter that whatever energy they put forth
was merely a recycling of the energy they took in through the food they ate.
Nevertheless, the energy embodied in the food may actually have a simul-
taneous combination of true investment and a recycling of already existing
energy.

It has long been understood that green plants use the chlorophyll molecule
to absorb sunlight and use its energy to synthesize carbohydrates (in this

case, sugars) from carbon dioxide and water. This process, known as photosynthesis (from the Greek *phos*, "light," plus *synthesis*, "to put together"), forms the basis for sustaining the life processes of all plants. The energy is derived from the sun (an original input) and combined with carbon dioxide and water (existing chemical compounds) to create a renewable source of usable energy. This process is analogous to an array of organic solar panels—the green plant.

Think of it this way: The plant (an array of solar panels) uses the green chlorophyll molecule (a "photoreceptor," from the Greek *phos*, "light," plus the Latin *recipere*, "taken back") to collect light from the sun within chloroplasts (small, enclosed structures in the plant that are analogous to individual solar panels). Then, through the process of photosynthesis, the sun's light is used to convert carbon dioxide and water to carbohydrates for use by the plant, a process that is comparable to converting the sun's light in solar panels on the roof of a building into electricity for our use. These carbohydrates, in turn, are partly stored energy from the sun—a new input of energy into the global ecosystem—and partly the storage of existing energy from the amalgam of carbon dioxide and water.[24]

When we eat green plants, the carbohydrates are converted through our bodily functions into different sorts of energy. By that is meant the energy embodied in green plants is altered through digestion into the various types of energy our bodies require for their physiological functions. The *excess* energy (that not required for physiological functions) is expended in the form of physical motion, such as energy to do work. On the other hand, it is different when eating meat because the animal has already used the sun's contribution to the energy matrix in its own bodily functions and its own physical acts of living, so all we get from eating flesh is recycled energy.

Biophysical Principle 4: All Systems Are Defined by Their Functional Performance

The behavior of a system—any system—depends on how its individual parts interact as functional components of the whole, not on what an isolated part is doing. The whole, in turn, can only be understood through the relationships, the interaction, of its parts. The only way anything can exist is encompassed in its interdependent relationship to everything else, which means an *isolated fragment or an independent variable can exist only as a figment of the human imagination.*

Put differently, the false assumption is that an independent variable of one's choosing can exist in a system of one's choice and that it will indeed act as an independent variable. In reality, all systems are interdependent and thus rely on their pieces to act in concert as a functioning whole. This being the case, no individual piece can stand on its own *and* simultaneously be part of an interactive system. Thus, *there neither is nor can there be an independent variable* in any system, be it biological (including the human thought process), biophysical, or mechanical because every system is interactive by its very definition—as a "system."

Moreover, every relationship is constantly adjusting itself to fit precisely into every neighboring relationship, each of which, in turn, is consequently adjusting itself to fit precisely into all its neighboring relationships. This universal dynamic *precludes* the existence of an independent variable because *no given thing can be held at a constant value beyond the number one* (the universal common denominator). To do so would necessitate detaching the thing in question from the system, which is a physical impossibility because all relationships are constituted by multiples of *one* in all its myriad forms, from quarks, atoms, molecules, and proteins, which comprise the building blocks of life, hence the living organisms themselves, which collectively form the species and communities. The only way the number one can exist, as the sole representative of any form on Earth, is to be the last living individual of a species, something intimated on the tribal level by James Fenimore Cooper's 1826 book *The Last of the Mohicans*,[25] because extinction is forever.

Therefore, to understand a system as a functional whole, we need to understand how it fits into the larger system of which it is a part and so gives us a view of systems supporting systems supporting systems supporting systems, ad infinitum.

Biophysical Principle 5: All Relationships Result in a Transfer of Energy

Although technically a *conduit* is a hollow tube of some sort, the term is used here to connote any system employed specifically for the transfer of energy from one place to another. Every living thing, from a virus to a bacterium, fungus, plant, insect, fish, amphibian, reptile, bird, mammal, and every cell in our body, is a conduit for the collection, transformation, absorption, storage, transfer, and expulsion of energy. In fact, the function of the entire biophysical system is tied up in the collection, transformation, absorption, storage, transfer, and expulsion of energy—one gigantic, energy-balancing act.

Biophysical Principle 6: All Relationships Are Self-Reinforcing Feedback Loops

Everything in the universe is connected to everything else in a cosmic web of interactive feedback loops, all entrained in self-reinforcing relationships that continually create novel, never-ending, irreversible stories of cause and effect, stories that began with the Eternal Mystery of the original story, the original cause. Everything, from a microbe to a galaxy, is defined by its ever-shifting interrelationship with every other component of the cosmos. Thus, "freedom" (perceived as the lack of constraints) is merely a continuum of fluid relativity. In contraposition, every relationship is the embodiment of myriad interactive constraints to the flow of energy, the very dynamic that perpetuates the relativity of freedom and thus of all relationships.

Hence, every change (no matter how minute or how grand) constitutes a systemic modification that produces novel outcomes. A feedback loop, in this sense, comprises a reciprocal relationship among countless bursts of energy moving through specific strands in the cosmic web that cause forever-new, compounding changes at either end of the strand, as well as every connecting strand. Here, we often face a dichotomy with respect to our human interests.

On the one hand, while all feedback loops are self-reinforcing, their effects in Nature are neutral because Nature is impartial with respect to consequences. We, on the other hand, have definite desires where outcomes are involved and thus assign a preconceived value to what we think of as the end result of Nature's biophysical feedback loops. To help you understand such a dichotomy, I (Chris) am going to tell you a story about salmon in the Columbia River Basin of the Pacific Northwest United States, which drains more than a quarter million square miles of land and is made up of 65 subbasins[26] with myriad streams of different orders. (*From the Forest to the Sea: The Ecology of Wood in Streams, Rivers, Estuaries, and Oceans* contains a comprehensive discussion of stream orders.[27]) Although the focus here is on salmon in the Pacific Northwest, the Atlantic salmon of the Eastern Seaboard has the same general life cycle (Figure 5.2).

A female salmon flexes her tail against the swift current as she propels herself to a small gravel bar just under the surface in the headwaters of a Pacific Coast stream. From somewhere in the shadow of trees overhanging the tiny stream, there comes a male; he arrives alongside the female with powerful undulations. They touch, and the female immediately turns on her side and fans the gravel with strong beats of her tail.

FIGURE 5.2
Atlantic Salmon. (Painting by Timothy Knepp, US Fish and Wildlife Service.)

She continues spraying gravel into the current until she creates a shallow depression, after which she begins depositing hundreds of reddish-orange eggs as the male squirts milky white sperm into the water. The cloud of sperm, enveloping the eggs, as the current carries it downstream fertilizes the eggs as they settle into the shallow "nest."

Now, she and her mate, having fulfilled the inner purpose of their lives, swim into deeper water, where they rest and die.

As the dead salmon wash into the shallow water along the edge of the stream's banks, the elements of their bodies become concentrations of nutrients and energy that feed the forest that helped nourish them as fertilized eggs. This massive infusion of decomposing salmon in the forested stream promotes the growth of algae and bacteria that help sustain aquatic insects.

As a carcass decomposes underwater, its dissolved nitrogen and carbon are soaked up by algae and diatoms, which are one-cell plants that form a scum on the gravel and rocks, which in turn are grazed by aquatic insects that will become food for the salmon that will hatch the next spring. In addition, the birds and mammals that feast on the carcasses, such as golden eagles, Steller's jays, ravens, wrens, spotted skunks, river otters, raccoons, bears, foxes, mice, and shrews, deposit their droppings on the forest floor.

Meanwhile, in the gravelly stream bottom lies an orange, opaque egg inside of which a salmon is developing. In time, the baby salmon hatches and struggles out of the gravel into the open water of protected, hidden places in the stream. Here, it grows until it is time to leave the stream of its origin and venture forth. It can go only one way—downstream to larger and larger streams and rivers until, at last, it reaches the ocean.

Juvenile salmon, steelhead, and cutthroat trout also poke around the expired, rotting bodies, eating the eggs left in the females and, eventually, picking off pieces of flesh. This huge addition of nutriments is critical for the young salmon because the rich banquet of dead fish enables youngsters to double their weight in about 6 weeks. The added body weight greatly increases the chances that a particular fish will survive to swim the gauntlet from the stream of its origin far out into the North Pacific Ocean and return again years later to spawn in the place it was hatched.

Only after some years at sea will the inner urge of individual salmon dictate their approaching time to spawn. Remembering the Earth's magnetic field, where they entered the ocean years earlier, they use it to navigate the open waters as they return to their home rivers, which they identify by the river's chemical signature. In so doing, salmon will differentiate into identifiable freshwater populations that are reproductively isolated from one another, each with its affinity to a particular river. Once in the river, they will again separate into discrete subpopulations, each with its own inner attraction to a particular stream within the river system, where it will spawn and die.

Nature's feat of nourishing the plants and animals requires about one salmon carcass per every 3 square feet of stream edge. This can be roughly translated into approximately one dead salmon for the amount of water that would today fill a standard bathtub.[28]

As for the self-reinforcing feedback loop, the greater the number of young salmon that survive to reach the ocean and die therein, the more the forest of their origin nourishes the ocean. In return, the more mature salmon that survive the ocean to reach their stream of origin and die therein, the more the ocean nourishes the forest.

Biophysical Principle 7: All Relationships Have One or More Trade-offs

For Biophysical Principle 7, we have a choice in everything we think and almost everything we do—except practicing relationships, experiencing ourselves as we experience relationships, choosing, changing the world, living without killing, and dying. In those, *we have no choice*, but *we do have a choice* of how we do it—and we must choose, in that *there is no choice*. Moreover, we must understand that not choosing *is still a new choice*, even if it appears to be doing nothing. In addition, we make a new choice each time a circumstance in our life changes, which, of course, is an ongoing process, be it the outworking of biophysical principles that govern life or how we view life's changes as we mature in years.

The constancy of change dictates the eternal presence of choice. Life can therefore be viewed as an eternal parade of decisions, each of which marks a fork in the path we follow. Each time a decision is made, others are foregone: They are the trade-offs. Nevertheless, each decision creates a kaleidoscope of additional choices that lead to either wisdom or folly, manifested in the ensuing consequences. In this sense, everything we think and do has a trade-off of hoped-for *or* unwanted consequences at the time a thought is formed, a decision is made, and the choice is executed.

Now, let us consider the Columbia River Basin and its 62 subbasins with the trade-off of its forest-to-ocean-to-forest feedback loop of salmon and other sea-going fish, such as steelhead (Figure 5.3). Although there are many dams on the Columbia River and its larger tributaries, the dam that has the most effect on the feedback loop is the Bonneville Dam, not far east of Portland, Oregon. This hydroelectric dam, which is closest to the Pacific Ocean, was built to supply electric power to parts of Oregon and Washington. The overriding collective trade-off is that the dam allows only an infinitesimal number of the salmon, steelhead, and other ocean-going fish, the quintessential part of the biological feedback loop in the entire Columbia River Basin, to reach their ancestral spawning streams, thereby preventing them from bringing ocean-based nutrients to the forest. Simultaneously, it prevents newly spawned fish from reaching the ocean, thereby preventing them from bringing to the ocean forest-based nutrients, among other things, such as driftwood, including centuries-old fallen trees.[29]

FIGURE 5.3
Map of the Columbia River drainage basin. (Map prepared by Kmusser.)

Biophysical Principle 8: Change Is a Process of Eternal Becoming

Change, as a universal constant, a unidirectional spiral, is a continual process of inexorable novelty. It is a condition along a continuum that may reach a momentary pinnacle of harmony within our senses. Then, the very process that created the harmony takes it away and replaces it with something else—always with something else. Change requires constancy, as its foil in order to exist, as a dynamic process of eternal becoming. Without constancy, change could neither exist nor be recognized.

When I close my eyes, I can again see the Saharan Desert of Egypt and feel the inner harmonious perfection of each sunrise and each sunset. I say "harmonious perfection" because for a half hour during the sunrise and a half hour during the sunset of each day, the temperature was perfect, the stillness beyond thought, and the vista immaculate in every direction.

Nevertheless, the eternal moment was ever-changing. By day, the sun stole my sense of perfection with its shimmering heat; by night, the infinity of space stole my sense of perfection with its gripping cold. Yet, even they had differing degrees of temperature within every minute I spent in the desert. For all things arise; all things reach a threshold between the appearance of stability and dramatic change; all things pass away.

Biophysical Principle 9: All Relationships Are Irreversible

Because change is a constant process orchestrated along the interactive web of universal relationships, it produces infinite novelty that precludes anything in the cosmos from ever being reversible. This being the case, we all cause some kind of irreversible change every day. To illustrate, I (Chris) remember a rather dramatic one I inadvertently made along a small stream flowing across a beach on its way to the sea. The stream, having eroded its way into the sand, created a small undercut that could not be seen from the top. Something captured my attention in the middle of the stream, so I unknowingly stepped on the overhang to get a better look, causing the bank to cave in and me to get a really close-up view of the water. As a consequence of my misstep, I had both altered the configuration of the bank and caused innumerable grains of sand to be washed back into the sea from whence they had come several years earlier while riding the crest of a storm wave.

One moment, I was standing on the level beach, and the next I was conversing with the water. At the same time, the sand I had knocked into the stream was being summarily carried off to the sea. What of this dynamic was reversible? Nothing was reversible because I could not go back in time and make a different decision of where to place my foot. And, because we cannot go back in time, nothing can be restored to its former condition. All we can ever do is repair something that is broken so it can continue to function, albeit differently from in its original form. If you want a detailed discussion of this principle, read *Earth in Our Care*.[30]

Biophysical Principle 10: All Systems Are Based on Composition, Structure, and Function

We perceive objects by means of their obvious structures or functions. Structure is the configuration of elements, parts, or constituents of something, be it simple or complex. The structure can be thought of as the organization, arrangement, or makeup of a thing. Function, on the other hand, is what a particular structure can do or allows to be done to it, with it, or through it.

Let us examine a common object, a chair. When did the notion of a chair begin to take form? This would be the first time in the far memory that one of our human ancestors sat on a piece of wood, which can be thought of as a "one-legged stool." From then on, our ancestors modified that piece of wood, both intellectually and practically, until stools with three or four legs were derived to fit various purposes, such as the old, close-to-the-ground, one-legged milking stool and today's tall, four-legged barstools. But, a stool is not necessarily restful. Then, somewhere in time, an ancestor had the idea of adding a back to a stool, and lo, the "chair" was born.

A chair is a chair because its structure gives it a particular shape *and* function. A chair can be characterized as a piece of furniture consisting of a seat, four legs, and a back; it is an object designed to accommodate a sitting person. If we add two arms, we have an *armchair*, wherein we can sit and rest our arms. Should we now decide to add two rockers to the bottom of the chair's legs, we have a *rocking chair*, in which we can sit, rest our arms, and rock back and forth while doing so. Nevertheless, it is the seat that allows us to sit in the chair, and it is the act of sitting, the functional component allowed by the structure, that makes a chair, a chair.

Suppose we remove the seat so the structure that supports our sitting no longer exists. Now to sit, we must sit on the ground between the legs of the once-chair. By definition, when we remove a chair's seat, we no longer have a chair because we have altered the structure and thereby altered its function. Now, if we leave the seat, but remove the back, the chair reverts to a stool. Thus, the structure of an object defines its function, and the function of an object defines its necessary structure. How might the interrelationship of structure and function work in Nature?

To maintain ecological functions means that one must maintain the characteristics of the ecosystem in such a way that its processes are sustainable. The characteristics one must be concerned with are (1) composition, (2) structure, (3) function, and (4) Nature's disturbance regimes that periodically alter an ecosystem's composition, structure, and function.

We can, for example, change the composition of an ecosystem, such as the kinds and arrangement of plants in a forest or grassland; this alteration means that composition is malleable to human desire and thus negotiable within the context of cause (manipulation) and effect (outcome). In this case, composition is the determiner of the structure and function in that composition is the cause, whereas the structure and resulting function are the effects.

Composition determines the structure, and structure determines the function. Thus, by negotiating the composition, we simultaneously negotiate both the structure and function. On the other hand, once the composition is in place, the structure and function are set, unless, of course, the composition is altered, at which time both the structure and the function are altered accordingly.

Returning momentarily to the chair analogy, suppose you have an armchair in which you can sit comfortably. What would happen if you either gained a lot of weight or lost a lot of weight but the size of the chair remained the same? If, on the one hand, you gained a lot of weight, you might no longer fit into your chair. On the other hand, if you lost much weight, the chair might be uncomfortably large. In the first case, you could alter the composition by removing the arms and thus be able to sit in the chair. In the second case, you might add a pillow or two or dismantle the chair, replace the large seat with a smaller one, and reassemble the chair.

In a similar but more complex fashion, the composition or kinds of plants and their age classes within a plant community create a certain structure that is characteristic of the plant community at any given age. It is the structure of the plant community that in turn creates and maintains certain functions. In addition, it is the composition, structure, and function of a plant community that determine what kinds of animals can live there, how many, and for how long. The animals, in general, are not just a reflection of the composition but ultimately are constrained by it.

If, for example, townspeople want a particular animal or group of animals within its urban growth boundary, let us say a rich diversity of summering birds and colorful butterflies to attract tourist dollars from birdwatchers and tourists in general, members of the community would have to work backward by determining what kind of function to create. With respect to birds, they would have to know what kind of structure to create, which means knowing what type of composition is necessary to produce the required habitat(s) for the birds the community wants. Thus, once the composition is ensconced, the structure and its attendant functions operate as an interactive unit in terms of the habitat required for the desired birds.

People and Nature are continually changing the structure and function of this ecosystem or that ecosystem by manipulating the composition of its plants, an act that subsequently changes the composition of the animals dependent on the structure and function of the resultant habitat. By altering the composition of plants within an ecosystem, people and Nature alter its structure and, in turn, affect how it functions, which in turn determines not only what kinds of individuals and how many can live there but also what uses humans can make out of the ecosystem.

Biophysical Principle 11: All Systems Have Cumulative Effects, Lag Periods, and Thresholds

Nature has intrinsic value only and so allows each component of an ecosystem to develop its biophysical structure, carry out its ecological function, and interact with other components through their evolved, interdependent processes and self-reinforcing feedback loops. No component is more or less important than another is; each may differ from the other in form, but all are complementary in function.

Our intellectual challenge is recognizing that no given factor can be singled out as the sole cause of anything. All things operate synergistically, as cumulative effects that exhibit a lag period before fully manifesting themselves. Cumulative effects, which encompass many little inherent novelties, cannot be understood statistically because ecological relationships are far more complex and far less predictable than our statistical models lead us to believe, a circumstance Francis Bacon may have been alluding to when he said, "The subtlety of Nature is greater many times over than the subtlety of the senses and understanding."[31] In essence, Bacon's observation recognizes that we live in the invisible present and thus cannot recognize ongoing cumulative effects.

The invisible present is our inability to stand at a given point in time and see the small, seemingly innocuous, effects of our actions as they accumulate over weeks, months, and years. Obviously, we can all sense change: day becoming night, night turning into day, a hot summer changing into a cold winter, and so on. But, some people who live for a long time in one place can see longer term events and remember the winter of the exceptionally deep snow or a summer of deadly heat.

Despite such a gift, it is a rare individual who can sense, with any degree of precision, the changes that occur over the decades of their lives. At this scale of time, we tend to think of the world as being in some sort of steady state (with the exception of technology), and we typically underestimate the degree to which change has occurred, such as the changes related to global warming. We are unable to sense slow changes directly, and we are even more limited in our abilities to interpret the relationships of cause and effect in these changes. Hence, the subtle processes that act quietly and unobtrusively over decades reside cloaked in the invisible present, such as gradual declines in habitat quality.

At length, however, cumulative effects, having gathered themselves below our level of conscious awareness, suddenly become visible. But, then, it is too late to retract our decisions and actions, even if the outcome they cause is decidedly negative with respect to our intentions. So, it is that cumulative effects from our activities multiply unnoticed until something in the environment shifts dramatically enough for us to see the outcome through casual observation. That shift is defined by a threshold of tolerance in the system, beyond which the system as we knew it suddenly, visibly, becomes something else. Within our world, the same dynamic takes place in a vast array of scales in all natural and artificial systems, from the infinitesimal to the gigantic.

A short-term example of cumulative effects, the lag period, and threshold is the cutting down of a neighbor's dying walnut tree. Initially, a man from the tree service sawed off the small branches with intact twigs. The effect was barely discernible at first, even as the twigs began to pile up on the ground. Each severed branch represented a cumulative effect that would have been all but unnoticeable had they not been accumulating under the tree.

After an hour or so (lag period) of removing the small limbs on one side of the tree, the cumulative effect gradually became visible as the number of limbs crossed the threshold. Had the same volume of twigs been removed throughout the tree and simultaneously gathered and removed from the ground, the cumulative effects would not have been as apparent. Nevertheless, the tree was gradually transformed into a stark skeleton of larger branches and the main trunk. Then, the large branches were cut off a section at a time, with the same visual effect as when the small ones had been removed, until only the trunk remained. The piecemeal removal of the tree created a slowly changing vista of the neighbor's house, until an unobstructed view of it appeared for the first time as another stark threshold was crossed.

If we now increase the spatial magnitude that encompasses the formation of a river's delta, the timescale involved for the cumulative effects to cross the threshold of visibility may well require centuries to millennia. When a river reaches the sea, it slows and drops its load of sediment. As the amount of sediment accrues on the seabed, it diverts the river's flow, causing it to deposit additional sediment loads in other areas (cumulative effects). Thus, over many years (lag period), the accumulated sediment begins to show above the water (threshold) and increasingly affects the river's flow as it forms a classic delta. The speed with which the delta grows has numerous variables, such as the amount of precipitation within the river's drainage basin in any given year, as well as the amount of its annual sediment load. Many of today's extant river deltas began developing around 8,500 years ago as the global level of the seas stabilized following the end of the last ice age.[32] So, the process of change and novelty continues unabated in all its myriad and astounding scales.

Biophysical Principle 12: All Systems Are Cyclical, But None Is a Perfect Circle

While all things in Nature are cyclical, no cycle is a perfect circle, despite such depictions in the scientific literature and textbooks. They are, instead, a coming together in time and space at a specific point, where one "end" of a cycle approximates—*but only approximates*—its "beginning" in a particular time and place. Between its beginning and its ending, a cycle can have any configuration of cosmic happenstance. Biophysical cycles can thus be likened to a coiled spring insofar as every coil approximates the curvature of its neighbor but always on a different spatial level (temporal level in Nature), thus never touching.

The size and relative flexibility of a metal spring determines how closely one coil approaches another: the small, flexible, coiled spring in a ballpoint pen juxtaposed to the large, stiff, coiled spring on the front axle of an 18-wheel truck. The smaller and more flexible a spring is, the closer are its coils, like the cycles of annual plants in a backyard garden or a mountain meadow. Conversely, the larger and more rigid a spring is, the more distant are its coils from one another, like the millennial cycles of Great Basin bristlecone pines growing on rocky slopes in the mountains of Nevada, where they are largely protected from fire, or a Norway spruce growing on a rocky promontory in the Alps of Switzerland.

Regardless of its size or flexibility, a spring's coils are forever reaching outward. With respect to Nature's biophysical cycles, they are forever moving

toward the next level of novelty in the creative process and so are perpetually embracing the uncertainty of future conditions, never to repeat the exact
outcome of an event as it once happened. This phenomenon occurs even in
times of relative climatic stability. Be that as it may, progressive global warming will only intensify the uncertainties.

In human terms, life is composed of rhythms or routines that follow the
cycles of the universe, from the minute to the infinite. We humans most commonly experience the nature of cycles in our pilgrimage through the days,
months, and years of our lives, wherein certain events are repetitive: day and
night, the waxing and waning of the moon, the march of the seasons, and the
coming and going of birthdays, all marking the circular passage we perceive
as time within the curvature of space. In addition to the visible manifestation of these repetitive cycles, Nature's biophysical processes are cyclical in
various scales of time and space, a phenomenon that means all relationships
are simultaneously cyclical in their outworking and forever novel in their
outcomes.

Some cycles revolve frequently enough to be well known in a person's
lifetime, like the winter solstice. Others are completed only in the collective
lifetimes of several generations, like the life cycle of a 3,000-year-old giant
sequoia in California's Sequoia National Park, hence the notion of the invisible present. Still others are so vast that their motion can only be assumed.
Yet, even they are not completely aloof because we are kept in touch with
them through our interrelatedness and interdependence with the one universal relationship.

Biophysical Principle 13: Systemic Change Is Based on Self-Organized Criticality

When dealing with scale (a small, mountain lake as opposed to the drainage basin of a large river, such as the Mississippi in the United States or the
Ganges in India), scientists have traditionally analyzed large, interactive systems in the same way that they have studied small, orderly systems, mainly
because their methods of study have proven so intellectually successful. The
prevailing wisdom has been that the behavior of a large, complicated system
could be predicted by studying its elements separately and by analyzing its
microscopic mechanisms individually: the *reductionist-mechanical mentality*
predominant in Western society that tends to view the world and all it contains through a lens of intellectual isolation. During the last few decades,
however, it has become increasingly clear that no large, complicated system,
be it a forest, an ocean, and even a city, yields to such traditional analysis.

Instead, large, complicated, interactive systems seem to evolve naturally
to a critical state in which even a minor event starts a chain reaction that
can affect any number of elements within the system and can lead to its dramatic alteration. Although such systems produce more minor events than
catastrophic ones, chain reactions of all sizes are an integral part of system

dynamics. According to the theory called "self-organized criticality," the mechanism that leads to minor events (analogous to the drop of a pin) is the same mechanism that leads to major events (analogous to an earthquake).[33] Not understanding this, analysts have typically blamed some rare set of circumstances (a perceived exception to the rule) or some powerful combination of mechanisms when catastrophe strikes.

Nevertheless, ecosystems move inevitably toward a critical state, one that alters the system in some dramatic way. This dynamic makes ecosystems dissipative structures in that energy is built up through time only to be released in a disturbance of some kind, such as a fire, flood, or landslide; in some scale, ranging from a freshet in a stream to the eruption of a volcano; after which energy begins building again toward the next release of pent-up energy somewhere in time.

Such disturbances, as ecologists think of these events, can be long term and chronic, such as large movements of soil that take place over hundreds of years (termed an *earth flow*), or acute, such as the crescendo of a volcanic eruption that sends a pyroclastic flow sweeping down its side at amazing speed. (A pyroclastic flow is a turbulent mixture of hot gas and fragments of rock, such as pumice, that is violently ejected from a fissure and moves with great speed down the side of a volcano. *Pyroclastic* is Greek for "fire-broken.")

Here, you might interject that neither a movement of soil nor a volcano is a living system in the classical sense. Although that is true, all disturbance regimes are part of the living systems they affect. Thus, interactive systems, from the habitat of a gnat to a tropical rain forest, perpetually organize themselves to a critical state wherein a minor event can start a chain reaction that leads to a catastrophic event—as far as living things are concerned— after which the system begins organizing itself toward the next critical state. Furthermore, such systems never reach a state of equilibrium, but rather evolve from one semistable state to another. This dynamic is precisely why sustainability is a moving target—not a fixed end point or a steady state.

Biophysical Principle 14: Dynamic Disequilibrium Rules All Systems

If change is a universal constant in which nothing is static, what is a natural state? In answering this question, it becomes apparent that the *balance of Nature*, in the classical sense (disturb Nature and Nature will return to its former state after the disturbance is removed), does not hold. In fact, the so-called balance of Nature is a romanticized figment of the human imagination, something we conjured to fit our snapshot image of the world in which we live. In reality, Nature exists in a continual state of ever-shifting disequilibrium, wherein ecosystems are entrained in the irreversible process of change and novelty, thereby altering their composition, interactive feedback loops, and thus the potential use of their available resources—irrespective of human influence. Perhaps the most outstanding evidence that an ecosystem is subject to constant change and disruption, rather than remaining in a static balance, comes from studies of naturally occurring external factors that dislocate ecosystems, and climate appears to be foremost among these factors.

After a fire, earthquake, volcanic eruption, flood, hurricane, or landslide, for example, a biological system may eventually be able to approximate what it was through resilience: the ability of the system to retain the integrity of its basic processes and overall relationships. But, regardless of how closely an ecosystem might approximate its former state following a disturbance, the existence of every ecosystem is a tenuous balancing act because every system is in a continual state of reorganization that occurs over various scales of time, from the cycle of an old forest to a geological phenomenon, such as Mona Loa, the active volcanic mountain in Hawaii.

Bear in mind that an old forest that is burned, blown over in a hurricane, or smashed in a tsunami could be replaced by another, albeit different, old forest on the same acreage. In this way, despite a repetitive disturbance regime, a forest ecosystem can remain a forest ecosystem. Thus, ancient forests around the world have been evolving from one critical, biophysical state to the next, from one natural catastrophe to the next. Whereas people can manipulate a forest to some extent, Mona Loa is entrained in an eternal flux of physical novelty over which no human has a smidgen of control.

Finally, we, the human component of the world, must understand and accept that the foregoing laws and biophysical principles are an interactive thread in the tapestry of the natural cultural world, which must be accounted for if society is to become a sustainable partner with—and within—a sustainable environment.

The upshot is that Nature's biophysical principles are the bedrock of social-environmental sustainability in all its myriad aspects throughout the world. It is therefore critical that the outcome of every environmental conflict so raises the level of the participants' consciousness that these principles are honored to the greatest extent humanly possible.

Discussion Questions

1. How does air affect the social-environmental sustainability of the world?

2. How is soil the nexus between living and nonliving components of life?

3. How does our behavior toward the land affect the oceans of the world and thus life?

4. How does our behavior toward the oceans of the world affect the land and thus life?

5. What is biodiversity? Why is it vital to the sustainability of life on Earth?

6. What is a "feedback loop"? How does a feedback loop affect biodiversity?

7. What is a biophysical principle?

8. Which are more important to the sustainability of life on Earth: Nature's biophysical principles or human-created laws?

9. Which is more important—the sustainability of Nature's biophysical processes or a desired economic outcome?

10. Since Nature's biophysical principles are inviolable, why not honor them?

11. Is one principle more important than another is? If so, which is the most important?

12. Can any principle operate independently from the others? If so, how? If not, why not?

13. How would life be different if Nature's biophysical principles were placed above human "rights" and thereby honored first and foremost in our social relationships with all of Nature?

14. How might honoring Nature's biophysical principles first and foremost in our social relationships with Nature affect the outcome of an environmental conflict?

15. Is there a particular question you would like to ask?

Endnotes

1. Ariella D'Andrea. Can the River Spirit Be a Person in the Eye of the Law? Global Water Forum, Governance Section. March 27, 2018. http://www.globalwaterforum.org/2018/03/27/can-the-river-spirit-be-a-person-in-the-eye-of-the-law/ (accessed July 15, 2018).

2. The foregoing discussion of being alone is from Chris Maser. *Of Paradoxes and Metaphors: Understanding Some of Life's Lessons.* Woven Strings, Amarillo, TX, 2008. 264 pp. E-book 254 kb.

3. Janice Chen, Yuan Chang Leong, Christopher J. Honey, and others. Shared Memories Reveal Shared Structure in Neural Activity across Individuals. *Nature Neuroscience,* 1(2017):115–125.

4. Ibid.

5. First Science.com. Carl Sagan. http://www.firstscience.com/home/poems-and-quotes/quotes/carl-sagan-quote_2284.html (accessed January 2, 2009).

6. Jonathan Rowe. The Hidden Commons. *Yes! A Journal of Positive Futures,* Summer (2001):12–17.

7. Chris Maser. *Forest Primeval: The Natural History of an Ancient Forest.* Sierra Club Books, San Francisco, 1989. 282 pp.

8. Krakatoa. http://en.wikipedia.org/wiki/Krakatoa (accessed March 13, 2010).
9. Chris Maser. *Interactions of Land, Ocean and Humans: A Global Perspective.* CRC Press, Boca Raton, FL, 2014. 308 pp.
10. Ibid.
11. Charudutt Mishra. *The Partners Principles for Community-Based Conservation.* Snow Leopard Trust, Seattle, WA, 2016. 180 pp. https://www.researchgate.net/publication/308888622_The_PARTNERS_Principles_for_Community-Based_Conservation (accessed March 3, 2018).
12. Theodor Seuss Geisel. *The Lorax.* Random House, New York, 1971. 56 pp.
13. Arnold J. Toynbee. *A Study of History, Volumes I–VI* (abridgement by D. C. Somervell). Oxford University Press, New York, 1987. 617 pp.
14. Russ Beaton and Chris Maser. *Economics and Ecology: United for a Sustainable World.* CRC Press, Boca Raton, FL, 2012. 191 pp.
15. (1) Rod Swenson. Emergent Evolution and the Global Attractor: The Evolutionary Epistemology of Entropy Production Maximization. In: P. Leddington, Ed., *Proceedings of the 33rd Annual Meeting of the International Society for the Systems Sciences,* 33(3), 1989, 46–53; and (2) Rod Swenson. Order, Evolution, and Natural Law: Fundamental Relations in Complex System Theory. In: C. Negoita, Ed., *Cybernetics and Applied Systems.* Dekker, New York, 1991, pp. 125–148.
16. Rod Swenson and Michael T. Turvey. Thermodynamic Reasons for Perception-Action Cycles. *Ecological Psychology,* 3(1991):317–348.
17. (1) Wolfgang Haber. Energy, Food, and Land—The Ecological Traps of Humankind. *Environmental Science and Pollution Research,* 14(2007):359–365; and (2) David Robson. Proto-Humans Mastered Fire 790,000 Years Ago. *ABC News,* October 28, 2008 (accessed February 27, 2009).
18. Ann Gibbons. Food for Thought. *Science,* 316(2007):1558–1560.
19. Rod Swenson. Spontaneous Order, Autocatakinetic Closure, and the Development of Space-Time. *Annals of the New York Academy of Sciences,* 901(2000):311–319.
20. BrainyQuote. Marcus Aurelius. http://www.brainyquote.com/quotes/authors/m/marcus_aurelius.html (accessed December 30, 2008).
21. G. O. Poinar, A. E. Treat, and R. V. Southeott. Mite Parasitism of Moths: Examples of Paleosymbiosis in Dominican Amber. *Experientia,* 47(1991):210–212.
22. The general discussion of amberization is based on (1) George O. Poinar, Jr. Insects in Amber. *Annual Review of Entomology,* 46(1993):145–159; (2) Scientist: Earth's Oldest Spider Web Discovered. London. *Corvallis Gazette-Times,* Corvallis, OR, December 16, 2008; and (3) Enrique Peñalver, David A. Grimaldi, and Xavier Delclòs. Early Cretaceous Spider Web with Its Prey. *Science,* 312(2006):1761.
23. Maser, *Interactions of Land.*
24. (1) Yuan-Chug Cheng and Graham R. Fleming. Dynamics of Light Harvesting in Photosynthesis. *Annual Review of Physical Chemistry,* 60(2009):241–262; and (2) Paul May. Chlorophyll. http://www.chm.bris.ac.uk/3motm/chlorophyll/chlorophyll_h.htm (accessed January 5, 2009).
25. James Fenimore Cooper. *The Last of the Mohicans.* Bantam Dell, a Division of Random House, New York, 2005. 416 pp.
26. Columbia River Basin Fish and Wildlife Program [click on basin map as GIF]. https://www.nwcouncil.org/fw/program/2000/2000-19/ (accessed February 23, 2018).
27. Chris Maser and James R. Sedell. *From the Forest to the Sea: The Ecology of Wood in Streams, Rivers, Estuaries, and Oceans.* St. Lucie Press, Delray Beach, FL. 1994. 200 pp.

28. The foregoing story of salmon is based on (1) Maser and Sedell, *From the Forest*; (2) C. Jeff Cederholm, David H. Johnson, Robert Bilby, and others. Pacific Salmon and Wildlife—Ecological Contexts, Relationships, and Implications for Management. In: David H. Johnson and Thomas A. O'Neil, Managing Directors, *Wildlife-Habitat Relationships in Oregon and Washington*. Oregon State University Press, Corvallis, 2001, pp. 628–684; (3) James R. Sedell, Joseph E. Yuska, and Robert W. Speaker. Study of Westside Fisheries in Olympic National Park, Washington. US Department of the Interior, National Park Service, Final Report CX-9000-0-E 081, 1983. 74 pp.; (4) J. M. Helfield and R. J. Naiman. Effects of Salmon-Derived Nitrogen on Riparian Forest Growth and Implications for Stream Productivity. *Ecology*, 82(2001):2403–2409; (5) Ellen Morris Bishop. Years of Adapting Separate Steelhead from Hatchery Cousins. *Corvallis Gazette-Times*, Corvallis, OR, March 5, 1998; (6) Timothy J. Beechie, George Pess, Paul Kennard, Robert E. Bilby, and Susan Bolton. Modeling Recovery Rates and Pathways for Woody Debris Recruitment in Northwestern Washington Streams. *North American Journal of Fisheries Management*, 20(2000):436–452; (7) James M. Helfield and Robert J. Naiman. Effects of Salmon-Derived Nitrogen on Riparian Forest Growth and Implications for Stream Productivity. *Ecology*, 82(2001):2403–2409; (8) Ted Gresh, Jim Lichatowich, and Peter Schoonmaker. An Estimation of Historic and Current Levels of Salmon Production in the Northeast Pacific Ecosystem. *Fisheries*, 25(2000):15–21; (9) Bruce P. Finney, Irene Gregory-Eaves, M.S.V. Douglas, and J. P. Smol. Fisheries Productivity in the Northeastern Pacific Ocean over the Past 2,200 Years. *Nature*, 416(2002):729–733; and (10) Nathan F. Putman, Kenneth J. Lohmann, Emily M. Putman, and others. Evidence for Geomagnetic Imprinting as a Homing Mechanism in Pacific Salmon. *Current Biology*, 23(2013):312–316. http://www.cell.com/current-biology/abstract/S0960-9822(13)00003-1 (accessed February 7, 2013).
29. Maser and Sedell, *From the Forest*.
30. Chris Maser. *Earth in Our Care: Ecology, Economy, and Sustainability*. Rutgers University Press, Piscataway, NJ, 2009. 262 pp.
31. Francis Bacon. http://Science.prodos.ORG (accessed January 2, 2009).
32. (1) Sid Perkins. O River Deltas, Where Art Thou? *Science News*, 172(2007):118; (2) Pippa L. Whitehouse, Mark B. Allen, and Glenn A. Milne. Glacial Isostatic Adjustment as a Control on Coastal Processes: An Example from the Siberian Arctic. *Geology*, 35(2007):747–750.
33. Per Bak and Kan Chen. Self-Organizing Criticality. *Scientific American*, January (1991):46–53.

6

The Human Equation

In any mediation process, people must be thought of and treated as though they are equal and deserving of love, trust, respect, and social-environmental justice, which asserts that we owe something to every person sharing the planet with us, both those present and those yet unborn. But, you may ask, what exactly do we have to give? The only things we have of value are the love, trust, respect, and wisdom gleaned from our experiences and embodied in each and every choice we pass forward. And, it is exactly because options embody all we have to give those living today and the children of tomorrow and beyond that social-environmental justice, as a concept, must fit within the context of human equality.

It must therefore be understood that the resolution of any environmental dispute will in some way affect the next generation—for good or ill. In this sense, a decision in the present has impacts in the future, and if they bode ill, it is analogous to taxation without representation. To the extent possible, each party must consider how a decision today might affect future generations.

If you wonder whether we even need to be concerned about the future, remember that adults are responsible for bringing children into the world. We are thus also responsible for being their voice and for protecting their options until the time when they can speak and act responsibly for themselves.

It has been our experience, however, that many people seem not to be overly concerned about the social-environmental circumstances future generations will inherit. I (Chris) have heard it asked, for example, "Why can't these changes wait until I've retired? Then someone else can worry about them."

Whether people are concerned about future generations depends on the way they grew up and the family values they learned: their initial social conditioning. How we, as children, learned to cope with circumstances influences how we treat one another as adults. Society is thus as peaceful or combative as we are as individuals.

The more a person is drawn toward peace and an optimistic view of life, the more functional (psychologically healthy) he or she is. The more a person is drawn toward debilitating, destructive conflict, cynicism, and pessimism about life, the more dysfunctional (psychologically unhealthy) that individual is.

To change anything in society, therefore, we must first look inward to confront, understand, and change ourselves, especially if we are going to act as a mediator in helping others to confront their fears and resolve their destructive conflicts. This process of self-evaluation and change puts the battle where it really belongs: within our own heart. As such, our inner struggles are the greatest learning experiences we will ever have. In addition, the greater our understanding of our own behavioral dynamics, of our own unresolved fears and pain, the easier it is for us to understand these dynamics in others and thus introduce compassion into the mediation of a conflict and wisdom into its visionary conclusion.

Facilitating the resolution of a conflict requires a basic understanding of those family dynamics that shape us as individuals. We therefore discuss familial dysfunction as openly and honestly as we can, which means that we must at times tap into our own familial dysfunction while growing up. This is important because to be a good mediator we must risk being an imperfect human. We must be open, honest, and vulnerable, and we must have the courage to work continuously and seriously on healing our own dysfunctions.

Our experience has been that we probably never really leave our families; we take them with us wherever we go throughout the rest of our lives. We take not only our families, both emotionally and intellectually, but also our familial heritage.

We sometimes make the assumption that our family experience is a common one, in that other families are just like ours. It is sometimes difficult to imagine because we only know our own experiences, our own perceptual "truths," just how different our individual "truths" can be. After all, *our perception is our truth*, which means everyone—*everyone*—is right from his or her point of view.

That said, however, we must always remember that, while we all take our families with us, it does not mean we have had a common or shared experience because each of us was raised differently, within a different culture, and with unique experiences, even within the same family. If we want to be successful at implementing environmental projects and policies, we must put aside our assumptions and learn to deal with *everyone*, as equally as humanly possible.

Thus, molded in the family template in an unknown and unknowable universe, the most consistently existential questions since the dawn of humanity have probably been, Who am I? and What value do I have in the immensity of the ever-changing Unknown and Unknowable? These are the fundamental questions we indirectly help the disputants address whenever we mediate the resolution of a conflict.

First, however, we must have some understanding of family dynamics because every conflict we have ever mediated was very much like stepping, as a stranger, into a family feud. Therefore, we begin sharing what we have learned about family dynamics by looking at the gift of a child, that little person whose psychic slate is initially untouched by socialized conditioning.

A Child's Gift

Childhood, the tender age during which we are taught to compete and fight, is the age in which the need for peacefully resolving conflicts is born. Thus, as a mediator, we must understand how life, which today may seem like a constant struggle to many people, replaces childhood innocence and creative possibilities through the social conditioning by parents, peers, schools, culture, and society.

To the Indian poet Rabindranath Tagore, "Every child comes with the message that God is not yet discouraged of man."[1] Each child offers their parents and society another chance to learn the meaning of love, to explore the boundaries of selflessness, to rediscover the possibilities of innocence, and to help us define who we are. Each child is a holy canvas on which we paint our loves and our fears, our joys and our sorrows, and a thousand other perceptions we hold to be ourselves.

Buckminster Fuller, American architect and inventor, thought that, "Every child is a genius but is enslaved by the misconceptions and self-doubts of the adult world and spends much of his or her life having to unlearn that perspective."[2] Thus, each child becomes the outward manifestation of the inner, adult self, for we see ourselves reflected in the children of the world. We, who are an imprint of our parental templates, have become the templates who will imprint ourselves on our children. And, what can we give them? "There are only two lasting bequests," according to Hodding Carter, "[that] we can hope to give our children. One of these is roots; the other wings."[3]

To this end, Dr. Edward Bach penned a beautiful paragraph on the ideal essence of parenthood:

> Parenthood is a sacred duty, temporary in its character and passing from generation to generation. It carries with it nothing but service and calls for no obligation in return from the young since they must be left free to develop in their own way and become as fitted as possible to fulfill the same office in but a few years' time.[4]

As children, we are molded in the template of our parentage, our peers, and our social environment, just as our parents were similarly molded: social conditioning. Too often, the result is that the precious gift with which each child comes into the world—innocence mirrored in spontaneous joy, aliveness, and creativity without preconceptions or limitations—is not recognized, not accepted, and deemed not acceptable.

Our innocence, which manifests itself as unbounded imagination, is stolen from us, often quietly and unobtrusively, through all sorts of external pressures to conform because we need to fit in rather than be something new, challenging, and exciting. Yet, it still seems to us that life is intended as a process of learning, a grand adventure, rather than a terror to be survived.

Nevertheless, however we turn out as adults, we are our family. And, unless we consciously choose otherwise, we take our family with us wherever we go, through our personal philosophy and through our behavior.

We Take Our Family with Us

Taking our family with us emotionally and psychologically is an important notion to understand, as a mediator, because we are inescapably our families to some degree. It is quite likely that the clearest things participants bring to the arena of conflict are their familial upbringing in terms of social conditioning. They are so entangled in their familial heritages they can seldom separate their dysfunctional behaviors from the ecological and social principles over which they fight. Helping the parties make this distinction is our task as mediators.

A word of caution is necessary here. While the behaviors exhibited by the disputants toward one another are often deemed condemnable, the person perpetrating the behavior is doing their very best at that moment and must be accepted with compassion—never condemned as a person.

It is not our intent to delve into a clinical discussion of family systems; excellent books on the subject are available. A brief overview of some dysfunctional familial dynamics is provided, however, because, as Mother Teresa said, "In the home begins the disruption of the peace of the world."[5] Understanding dysfunctional familial dynamics is therefore critical to good mediation.

Dysfunctional Family Dynamics Lead to Ongoing Destructive Conflict

Dysfunctional behavior often leads to conflict, and thus it is absolutely necessary for us, as mediators, to understand dysfunction because the more dysfunctional a person is, the more inclined the person is toward conflict. Mediation is therefore a process whereby we help the parties consciously break their dysfunctional cycle of destructive behavior so they can resolve their conflict. To be a good mediator, however, we must work seriously on resolving *our own dysfunctional behaviors*.

Only when we are free of our own dysfunctional familial patterns can we really be open to the humility, spontaneity, and creativity demanded by the mediation process. Only then can we offer the understanding, insight, and empathy necessary to lead and communicate effectively. Each person's story is the same in principle, but differs in detail.

First, know thyself. Mediation is a skill that is critical to our success as environmental leaders and caretakers. But, how are we to learn to help others if we are unaware of our own dysfunctional behaviors?

Learning about your personality is a good start, but discerning how you are perceived by others and how to modify your own dysfunctional behavior is much more valuable to you as a mediator. If, for instance, you find yourself in conflicts that leave you wondering how you got there, then we strongly suggest that you consider furthering your personal growth. Becoming self-aware is one way to grow beyond the child and family situation of your past and become the psychologically mature adult you truly wish to be.

We are the strengths and the weaknesses of our upbringings because we all go through similar dynamics in various forms as we come into, grow in, and leave our families. We thus tend to repeat the patterns—whether they are functional or dysfunctional—over and over again unless we consciously break an unwanted cycle. To break a dysfunctional cycle, one must first understand homeostasis.

Homeostasis Is Designed to Hide Dysfunction

It is critical that we, as mediators, understand homeostasis and homeostatic mechanisms because, just as each dysfunctional family has its own set of mechanisms, so each party within a given conflict has its own. And, a party's mindset is fashioned through an unconscious amalgamation of each person's familial pattern consolidated into a collective pattern— that of the disputing participants. (The term *homeostasis* is from the Greek *hómoios*, "similar," and *stasis*, "stand still.")

As such, homeostasis is the maintenance of a dynamic equilibrium within a system, such as a family. A family is a system governed by a set of rules that determine and control the interaction of its members in organized, established patterns. The family rules are a set of directives concerning what shall and shall not occur within and outside the family. Homeostatic mechanisms maintain the ongoing arrangement among family members by activating the rules defining each member's relationship to the whole.

My (Chris's) father, for instance, could not control the imperfections in his family no matter how hard he tried because they were really the imperfections he perceived within himself, which he transferred or projected onto us. In turn, his perception of our imperfections triggered his abusive behavior, and his abusive behavior was the secret skeleton in our family closet.

My family probably appeared to be quite "normal" on the surface. Seen through knowledgeable eyes, however, which could have interpreted the symptoms I acted out as a child away from home, the red flag of abuse would have been readily apparent.

Dysfunction and homeostasis are therefore self-perpetuating, self-reinforcing feedback loops founded on coercion and fear. My father was abusive, and that would have drawn criticism. To avoid the criticism, we were all assigned roles to play, which kept the dynamic equilibrium in the family within acceptable bounds; this in turn kept the dysfunction within the family while giving the outward appearance of normalcy.

The roles we were assigned were to be the perfect son, daughter, wife, and mother—according to my father's definitions. This was particularly important where his public image was concerned. If the homeostasis began to crumble, so would his perception of other people's perceptions of the family image, of his image, and that was an unacceptable threat to my father's sense of survival. And yet, *he did his best under the circumstances of his being unwanted and thus abandoned twice at an early age and his subsequent upbringing in an orphanage* until the age of 15, when he was finally adopted.

At this same juncture stand the disputing parties in every environmental conflict, feeling an unacceptable threat to their survival. As prisoners of their familial upbringing, they are searching, albeit often unknowingly, for appropriate behavioral boundaries within which to feel safe.

Boundaries, the Silent Language

Boundaries are those lines of silent language that allow a person to communicate with others while simultaneously protecting the integrity of one's own personal space, as well as the personal spaces of those with whom one interacts.

The language of boundaries transcends individual space to include familial space, cultural space, and even national space. Understanding personal boundaries during mediation among individuals of the same culture is difficult enough, but expanding that concept into a fluid working ability among different cultures is most difficult to accomplish. This is especially true in other countries, where mediation may be done through a translator in a language one can neither understand nor speak.

A simple way of looking at boundaries is the adage "good fences make good neighbors" from Robert Frost's poem "Mending Wall."[6] As an example, consider cliff swallows, which attach their mud nests to such surfaces as the faces of cliffs, the sides of buildings, and under bridges. These enclosed, globular nests share common walls, which not only strengthen the nests but also keep the peace by preventing the inhabitants from peeking into each other's abodes. If, however, a hole is made in the common wall and the swallows can see each other, they bicker and squabble until the hole is repaired, which immediately restores tranquility.

A more complicated way of dealing with physical boundaries is to compare them to the home ranges and territories of animals. A home range is that area of an animal's habitat in which it ranges freely throughout the course of its normal activity and in which it is free to mingle with others of its own kind. A territory, in contrast, is that part of an animal's home range that it defends, for whatever reason, against others of its own kind. This defensive behavior is most exaggerated and noticeable during an animal's breeding season.

How does this concept apply to us? Suppose it is Saturday morning and you leave your home to go take care of a few errands. You simply go about your business without paying much attention to what is going on around you or to the people you pass unless you happen to meet someone you know. In general, you are simply engrossed in what you are doing. When you have finished your errands, you start home.

The closer you get to your neighborhood, the more alert you unconsciously become to changes around you, such as the new people moving in two blocks away. This "protective feeling" becomes even more acute as you approach the area of your own home and notice a car with an out-of-state license plate parked in your neighbor's driveway. You get out of your car and immediately notice, perhaps with some irritation, that the neighbor's dog has visited your lawn again while you were gone. If your neighbor's dog had anointed someone else's yard with its leavings, you probably would have paid no attention.

The same general pattern extends into your home. Inside your home, how well you know someone and how comfortable you feel around them determines the freedom with which they may interact with you and your family and use your house. You are the most particular about your ultimate private space, your physical being.

For example, an unwanted salesperson may not be allowed inside your home. A casual acquaintance, on the other hand, may be allowed in the living room and use of the guest bathroom but is not allowed to wander about the house without permission. If one of your children's friends comes over, they may be allowed into the living room, kitchen, family room, guest bathroom, and your children's rooms (but only with permission), but they are not allowed into your bedroom. At times, even your children may not be allowed in your room without your permission, or perhaps you in theirs, as they mature.

Although this dynamic may function in a "normal" manner for strangers, it often becomes so blurred among the members of a dysfunctional family, wherein personal boundaries, including the physical body itself, are violated. In some families, appropriate personal boundaries are all but absent. This dysfunctional trait is usually carried into the arena of environmental conflict.

It is, therefore, our task to set the behavioral boundaries, as rules of conduct, which not only exist as the infrastructure of society but also make the mediation process work, and to make sure that all participants understand and respect these boundaries. Understanding and respecting boundaries helps to build and maintain trust. This is critically important because interpersonal boundaries are an absolute social necessity of communication.

Let us look at a few concrete examples, beginning with the mediator. The most important interpersonal boundary for us, as mediators, to maintain is that of a guest at all times because we serve at the participants' behest. By staying within "guest boundaries," we are nonthreatening and can create and maintain a safe environment within which the participants can struggle to communicate. In a tense setting, humor can lighten the atmosphere. So, to maintain this safe space, I (Chris) typically utilize humor in the form of puns, which are simply plays on words, rather than other kinds of jokes that might offend.

One of the more important behavioral contracts that we make with the participants is to listen to one another without interrupting. If necessary, a "talking stick" is passed from person to person as they speak. The talking stick signifies the right of the holder to speak and the obligation of everyone else to listen in silence. This is imperative because waiting one's turn is part and parcel of civility and equality, both of which are prerequisites for a safe environment.

In our experience, ground rules are best developed and agreed on by the group. This agreement not only is important in maintaining order but also is one of the first decisions that group will make together. As such, it is important for us, as mediators, to recognize that the process of establishing ground rules may set the tone for all future group interactions. The following are examples of ground rules:

1. Meetings begin and end on time.
2. One person speaks at a time (the talking stick example).
3. Listen respectfully when someone else is speaking.
4. Speak only for yourself.
5. Understand that everyone is "right" from their point of view; thus, viewpoints are "right, right, and different," with difference being negotiable.
6. Value and respect all perspectives.
7. Treat everyone with respect.
8. Make no personal attacks.
9. Discuss issues, not positions.
10. Instead of dwelling on past, failed efforts, focus on working toward a positive future.
11. If uncomfortable with a decision or direction, speak up.

These are merely examples, but it is easy to see how ground rules set the tone for the discussions and negotiations to come.

Although setting ground rules may sound fairly simple, learning to understand boundaries is often complicated by the various "coping mechanisms" through which we, as children, learn to survive and with which disputants attempt to quell their fears while dealing with one another. These mechanisms comprise a part of everyone's personality characteristics and, as such, are important for us to understand because they are often the key to unlocking the stubborn discord of conflict.

Coping Mechanisms: Unconscious Thoughts that Manifest as Recognizable Behaviors

Coping mechanisms, first deciphered and named "defense mechanisms" by Sigmund Freud, who developed psychoanalysis, begin as thought processes we devise to protect ourselves from that which we deem dangerous to our well-being. What begins as a thought manifests as a behavior when we are confronted with the perceived life-threatening circumstance from which the thought process was originally devised to protect us. If the combination of thought and action is successful, then we have devised a functional mechanism of survival, a "coping mechanism," which increasingly becomes a self-reinforcing feedback loop every time it works as we expect it to work. As we begin to use it automatically, the thought process is relegated to our subconscious, and only the behavioral pattern is manifested.

Coping mechanisms therefore become the unconscious behavioral devices we learn to use to help us retain or regain control in uncomfortable situations. This really means we are trying to cope with a universe in constant change.

Coping mechanisms, as a strategy for survival, are often functional, positive, and entirely appropriate for a given circumstance when we develop them, but they eventually can, and often do, become outmoded and dysfunctional as circumstances change. Clinging to dysfunctional coping mechanisms when they fail to meet current or new situations in life can lead to a hardening of attitudes, a "hardening of the heart" so to speak, and create a rigidity that leads to destructive conflict.

Because dysfunctional coping mechanisms involve self-deception and a distortion of reality, they do not resolve problems; they only alleviate symptoms. Moreover, because they operate on a relatively unconscious level, they are not subject to the normal checks and balances of conscious awareness.

Coping mechanisms—we all have them. Although there is a vast array of coping mechanisms, only a few of the more common ones are discussed here. It is important that we, as individuals, consider those coping mechanisms that we use most and make an effort to recognize them—and control them—before they jeopardize our efforts. James Tamm and Ronald Luyet provided a list of 50 signs of defensiveness in their book, *Radical Collaboration.*[7] According to Tamm and Luyet, "Defensiveness does not defend [one] from others but arises to protect us from experiencing our own uncomfortable feelings."[8] Recognizing that we share the same or similar coping mechanisms is just one more reminder that we are all human and, as such, do the best we can at all times.

Keep in mind that the point of this discussion is *not* that coping mechanisms are bad or that we need to rid ourselves of our particular tactics for coping with life. Coping mechanisms are not clearly defined, separable behavioral patterns, but rather are overlapping behaviors, which grade into and out of one another almost at will. It is thus the awareness, the consciousness, with which we observe our behavior and that of others that is at issue. Therefore, we must ask ourselves with compassion and forthrightness, Does this behavior and its underlying motivation best serve my present needs in life?

We do not dwell here on the coping mechanisms per se, but rather give brief examples to illustrate how they are used because they form the backbone of conflict's thrust and parry. As such, it is absolutely necessary for us to understand coping mechanisms because the more functional a person is, the more inclined they are toward peace; conversely, the more dysfunctional a person is, the more inclined they are toward conflict.

A few years back, I (Lynette) participated in workshops on environmental management. On the first day, there were exercises to build leadership and facilitate team bonding. One particular icebreaker activity involved randomly going up to people in the group and, through quick conversation, determine what common interests we shared. Moving around the room, I had a conversation with someone I was somewhat acquainted with, at least from my perspective. Nevertheless, our communication always seemed slighted tainted (from previous interactions) and always slightly stuck. So, my expectations going into this part of the activity were rather low. At first, we hit a roadblock; we went through a list of our individual interests with no match. Through further conversation, we discovered that we had a special bond. Both of us were trained as lifeguards. I was moved. Something about making that connection altered my thought pattern. And, even today, though my communication with that individual has not substantially changed, I am more willing to overlook those factors and focus on our unique bond. This exercise reframed my thought process regarding the individual, far beyond the workshop time frame. This is an indication to me that, if I choose, my shift in perception can positively change my interactions.

Recognizing and understanding the language of coping mechanisms, especially dysfunctional ones, are vital in understanding a person's family dynamics and how that person deals with life. It is thus a critical step in understanding the dynamics of a conflict and how to help the disputants resolve their differences with dignity. As you read the following descriptions, you might think of examples from your own life and family upbringing.

Anger and Aggression

Anger and aggression are discussed together because anger is the emotion that triggers aggression as the act. Anger is a feeling of extreme displeasure, hostility, indignation, or exasperation toward someone or something. Anger is extreme fear or frustration violently projected outward and is synonymous with being upset, feeling a minor irritation, or feeling an intense rage. It is a temporary insanity that isolates us from the facts, from ourselves, and from one another.

In addition, we (you and I) are not angry for the reason or at the person or thing we think we are. We are always angry with ourselves for being afraid of circumstances and therefore feeling out of control, which has nothing to do with the person or thing at which we direct our anger.

Unfortunately, this realization all too often follows my (Chris's) anger, which I have attempted to project onto someone or something else. I also find that a minor irritation is of the same category as intense rage because the dynamic is the same; it is only a less intense *reaction* on the same continuum. I feel internal disharmony, which is fear of a circumstance in which I feel a loss of control, and I am angry about feeling afraid of the circumstance.

Unless we (you and I) fully understand this dynamic, we think we are really angry for the reason or with the person or thing at which we level our anger. We therefore use our anger as a means of *not* having to deal with the circumstance that we are really afraid of. This often happens at meetings between agency representatives and the public. The latter often hear things with which they disagree and over which they have no control. Instead of listening calmly, they become angry and start yelling. I (Chris) have, in fact, mediated meetings where participants have been so charged with emotion that they became red in the face, yelled, cursed, and physically shook with rage.

The Golden Rule must always apply in situations dominated by anger and aggression. Treat others in the way you would like to be treated. An angry coal miner was probably afraid he would be denied his permit and so lose something most dear to him: his livelihood. Instead, I was able to deflect his anger and show him respect in spite of his outburst. As a result, he sat down at the table and became a perfect gentleman for the remaining time we had together.

In the intensity of the emotion, people often feel that they are right in projecting their anger onto those who seem to be in control, those who have "taken" control away from them. In the grip of their anger, they do not perceive that they have a choice because they feel a loss of control, which they find terrifying.

Anger usually translates into aggression, which, as it is used here, is the habit of launching attacks, of being hostile. If we (you and I) show enough aggression toward a person or persons with whom we think we are angry, then we are coping with our fear by "reacting" to the particular circumstance, which causes the person or persons to back away from the threatening energy. Through aggression, we think we can avoid having to deal with the circumstance over which we have no control and of which we are frightened.

All we have really accomplished, however, is to isolate ourselves from any understanding of the data and from the people who are presenting the data. If, on the other hand, we "respond" to the circumstance by being patient and open-minded, and gently asking questions, then we might be able to overcome our fear and in so doing realize that there are no enemies out there, only other people who, like us, are frightened of the unknown.

Appraisal

Appraisal is the act of evaluating something; of estimating its quality, amount, size, and other features; of judging its merits. As such, appraisal is an interesting coping mechanism in that it can effectively prevent forward motion. It is like being the traveler on the platform at the train station who is so afraid of missing the train that they spend all their time checking and rechecking the schedule, being so preoccupied evaluating the schedule, they do not even see the train come and go.

Another example of an overappraiser is the shopper who goes to the grocery store to buy three items and has to read every comparative label in minute detail and then weigh and reweigh the data before making a choice. Thus, what would take most people 5 minutes to buy takes such an appraiser 45 minutes.

Appraisers cope with their fear of criticism by checking, rechecking, and further rechecking the data, and they are seldom willing to make a decision for which they may be held accountable. When in doubt, they conduct another study, but refrain, at any cost, from *saying* or *doing* anything until all the data are collected and have been carefully and properly analyzed. This, of course, will never happen because even if all the data could be collected, the appraiser would still continue the analysis indefinitely because there is no such thing as "definitive" data.

Over the years, I (Chris) have met many people who appraise their life away. Although I find them exceedingly difficult to work with when decisions need to be made, I feel deep compassion for them because I remember times in my boyhood when I also was too terrified to make a decision, knowing I would be humiliated, beaten, or both no matter what I did.

Defensiveness

To be defensive means to protect that which already is; to resist a new view, to resist the possibility of change, and to resist the truth about oneself. Defensiveness limits your growth in that you argue for your old self rather than taking a new look and embracing a new possibility. You defend the rut in which your old belief, your old behavioral pattern is stuck. You become defensive because at some level you know that what is being said is at least partly true, and if you acknowledge that truth, you will have to act on it, which means changing your stance. Thus, you feel obliged to defend the old groove. After all, it is a comfortable, known entity, like home.

Defense as a coping mechanism takes planning. A planned life can perhaps be tolerated but cannot be fully lived.

> The mind engaged in planning for itself is occupied in setting up control of future happenings. It does not think that it will be provided for, unless it makes its own provisions. Time becomes a future emphasis, to be controlled by learning and experience obtained from past events and previous beliefs. It overlooks the present, for it rests on the idea the past has taught enough to let the mind direct its future course.
>
> The mind that plans is thus refusing to allow for change. What it has learned before becomes the basis for its future goals. Its past experience directs its choice of what will happen. And it does not see that here and now is everything it needs to guarantee a future quite unlike the past, without a continuity of any old ideas and sick beliefs. Anticipation plays no part at all, for present confidence directs the way.
>
> Defenses are the plans you undertake to make against the truth. Their aim is to select what you approve, and disregard what you consider incompatible with your beliefs of your reality. Yet what remains is meaningless indeed. For it is your reality that is the "threat" which your defenses would attack, obscure, and take apart and crucify.[9]

Denial

Denial is refusal to recognize the truth of a situation; it is a contradiction, a rejection of what is. Although denial, as a coping mechanism, is part and parcel of almost all other coping mechanisms, it is also an entity unto itself.

Think, for example, of your mind as the honeycomb in a beehive and visualize yourself stuffing your unwanted feelings into an empty comb and sealing it so you will not have to deal with them: out of sight, out of mind. You are now effectively in denial of your feelings. The rest of your mind seems to be cleared of your discomfort. You are free to live, but only as long as you can continually mend the combs already filled and continually create more combs to accommodate future discomfort.

Denial is one of the most pervasive coping mechanisms in the world. It is such a simple device that it is probably the great-great-grandparent of all coping mechanisms. The following is a typical example of denial. Speaking intently and quickly, the dark-haired, green-eyed young woman explained she had been sexually abused by a relative from the time she was 5 years old until she was 14. Although she had two miscarriages, her parents still refused to believe her, which means they were denying that anything improper had taken place.[10] And, today there is the ongoing denial of global warming by various government officials, as well as corporate members of the oil, gas, and coal industry, despite all the photographs and on-the-ground measurements that unequivocally verify the worldwide melting glaciers and the Arctic, as a whole, as well as the global rise in level of oceans.

Displacement

Displacement, as a coping mechanism, is used to shift the focus from that which is uncomfortable to that which is safe; it is often referred to as a "smoke screen." I (Chris) have had attorneys for the federal government try to distract me with this tactic while I was under oath as an expert witness. They did not want me to complete my answer to a question they had asked because they were afraid of what I was saying, so they interrupted and asked a totally unrelated question. Recognizing this tactic, however, I always completed my answer to the first question and then answered the displacement question.

Another way to cope with fear of losing control is to displace the real reason with the use of time. Some people have their lives so tightly scheduled that they *do not have a minute to waste.* They confuse motion and time constraints with productivity. In this way, they control what they do, who they see, and how long they see them without ever having to take responsibility for saying: "I don't want to see you because you make me uncomfortable," or "I don't want to see so and so because I'm afraid that I might fail, which I cannot handle right now."

People therefore use time to control those circumstances they wish to deal with and to see those people they choose to see for as long as they want to see them. At the same time, they are pleading a case for being innocently out of time—out of control—for those circumstances or those people they do not want to deal with.

One retired man I (Chris) knew was so afraid of dying that he displaced his fear onto time; consequently, he had no time to waste. The sad thing was that he was so conscious of his "time running out," as he put it, that he did not enjoy what he did because he never had time to do it. He always had to get on to the next thing. He raced time around his prison cell—his fear of dying— and literally wore himself out before his *time* might have been up. His coping mechanism had become his unconscious agent of indirect suicide.

Filters

A filter is a device through which a substance (such as light, water, or thought) is passed to remove "unwanted impurities." In the sense of a coping mechanism, people filter out unwanted material as a way to *accept* and *understand* whatever they want to accept and understand. For example, have you ever tried explaining something to someone and had them hear only part of it, the part they *wanted* to hear?

I (Chris) have often found this to be the case when I spoke to a group of people comprising the timber industry, environmental organizations, and land management agencies. They each heard what they wanted to hear in what I said, and they each addressed these different aspects of the presentation during the question-and-answer period. The more polarized an audience is, the more predictable are the questions they are likely to ask and the responses they are likely to hear and accept.

At times, people live as though they are in a giant "safe" with filters to control what they see, what they hear, and what they feel. They thus hear only what they want to hear, see only what they want to see, and feel only what they want to feel. They can accept and understand that which they choose but do not have to move out of their comfort zone and be accountable in the world. This is what it means to "look at life through rose-colored glasses." Filtering is thus a common coping mechanism to "selectively" hear and see, as exemplified in two of the three monkeys: hear no evil and see no evil.

Filters can be very frustrating for the person who is trying to communicate with someone who does not want to hear what is being said. Yet, we all filter information simply because we have different frames of reference. Speaking for myself, I (Chris) always endeavor—and often fail—to lay aside my filters so I may become educated in the sense that the poet Robert Frost meant when he wrote, "Education is the ability to listen to almost anything without losing your temper or your self-confidence."[11]

Projection

Projection is a casting forward or outward of something. As a coping mechanism, it means the externalization of an inner thought or motive and its subsequent behavior, which is then attributed to someone else.

> And Aaron shall lay both his hands upon the head of the live goat, and confess over him all the iniquities of the people of Israel, and all their transgressions, and all their sins; and he shall put them upon the head of the goat, and send him away into the wilderness.... . The goat shall bear all their iniquities upon him to a solitary land; and he shall let the goat go in the wilderness.[12]

In biblical times, on the Jewish day of atonement, Yom Kippur, all the transgressions of the Jewish people were heaped (projected) onto the back of a "scapegoat," which was then driven away into the wilderness, "taking" all the people's transgressions with it. Thus, projection as a coping mechanism has a long-recognized history.

Just as an empty movie projector casts only light, you can project onto other people only what you think about yourself because without thought, there is nothing to project. Thus, you see in others what you both consciously and unconsciously see in yourself, nothing more, nothing less.

As such, judgment, the projection of that which you see in yourself, is the projectile you cast outward in the word *should*: you *should* do this; you *should* do that. "You should" is thus a common attitude of the opposing sides in a conflict. The opposite is equally someone else's concept of a social misconduct: You "shouldn't" do this or that.

In reality, however, should and shouldn't are both the stuff of someone else's standard of operation, of someone else's concept of right and wrong, of what you should or should not be or do. Someone else's should or shouldn't is yours only if you choose to accept it. On the other hand, you can choose to ignore another person's admonishment, and then it has no effect.

Projection is a very common coping mechanism. When understood as such, projections can be enlightening. You can, for example, tell almost immediately how participants feel about themselves by the kinds of projections they level at their opponents.

Rationalization

To rationalize in the sense of a coping mechanism is to devise self-satisfying, but inauthentic, reasons for one's own behavior. For example, you have been told to do something in your job with which you do not ethically agree. If you do not comply, however, you will lose your job, a real possibility in these days of corporate/political administrations. Therefore, you rationalize that you can do more good working for change on the inside of the agency or company by compromising your beliefs than you can by getting yourself fired for sticking to your principles. In so doing, you intellectually rationalize acceptance of the order and comply with it, but you have simultaneously committed the ethical honesty of your feelings to the prison of repression. Thus, you have murdered a vital, creative part of yourself.

The most commonly used rationalization is lack of control: "I can't." *Can't* means that whatever it is, it is beyond your control. Therefore, you are not responsible for your behavior. What you are really saying when you say "can't" is, "I will not, I choose not to, I am afraid to" or some similar declaration.

This rationalization probably came about because not knowing the answer to a question is not acceptable in our society. "I don't know" is reinforced as an unacceptable answer from grade school through college, the military, the workplace, and life in general. In fact, when you think about it, the statement, "Ignorance is no excuse under the law," is saying the same thing. Not knowing is not okay.

Repression

Repression can be thought of as a one-way, spring-loaded valve into your unconscious. Any thought or emotion that causes you anxiety passes through this one-way valve, building tension in the coiled spring as it does so. Once trapped in the unconscious, neither the thought nor the emotion is allowed to reappear in your awareness. It might be expressed as follows: Homer really wanted to slug his brother for having offended Alice, but that would not have been acceptable at the party. So, he clamped his jaws together and clenched his fists as he stalked from the room, tamping down his anger— putting a lid on it—so it would not erupt unacceptably. He repressed his emotions. Without an acceptable "safety valve" for releasing tension, energy continually builds in the spring until Homer will one day "blow up" unexpectedly and badly hurt his brother over some trivial matter.

I (Chris) have seen this coping mechanism in agency personnel. On being ordered to do things for the good of the agency, things that violated their moral sense of what was right, they repressed their emotions to keep their jobs rather than maintain the integrity of their beliefs, even if it meant resigning. The moment they retired, however, they attacked the agency with all the pent-up vehemence and bitterness of those long-repressed emotions.

Resistance

To resist is to work against, to fight off, to actively oppose. Resistance is simply a conservative, stabilizing tendency that keeps an individual from overstepping limitations too quickly and rashly.

Problems arise when your resistance to change becomes overreactive, obsolete, or maladaptive, in a word, dysfunctional. Then, you are unable to express your potential or to meet your goals. Resistance, in the dysfunctional sense, is one of the most commonly used coping mechanisms to ward off change, to avoid the responsibility of moving forward, of participating in life.

Resistance is like swimming directly against the current of a large, swift river. A swimmer in such circumstances becomes worn out, despite maximum effort, and is carried downriver by the overwhelming, persistent strength of the current and sometimes drowns. If, perchance, the swimmer is strong enough and determined enough just to stay even with the current, it soon becomes apparent that, while the current does not tire from its flowing, the swimmer tires from the effort of swimming. Thus, the current ultimately carries the swimmer away: the tired carried away by the tireless.

Circumstances are the river of life, and change is its current. The individual swimmer can choose to resist the current, become fatigued, and perhaps drown or can choose to flow with the current and, with patience, learn the skill necessary to cross the river safely and easily. Herein lies the secret of control: To be in control, you must give up trying to control. Only when you give up trying to control life can you master navigating its current.

Today, the pace at which change is occurring is unprecedented in our history. More and more, we feel like we have lost control, and this sense of loss often results in repressed anger and resistance.

That which you resist persists in the degree to which you resist it, and you become like that which you resist. It cannot be otherwise. What you resist is a lesson in life not learned, and life seems to persist in its lessons until you learn them.

We find, however, that resistance serves two purposes in life, one positive and one negative. A feeling of resistance is positive when it is your inner voice—the voice of your heart—telling you that what you have been asked to do really goes against your deepest sense of principles. In this case, you can feel good honoring your resistance.

On the other hand, there are times when you simply do not want to do something that needs doing. Then, your resistance works against you. You may end up with a terrible headache because your resistance is like driving a car by stepping on the gas pedal and the brakes at the same time with equal pressure.

Standards and Judgment

A standard is an acknowledged measure of comparison for qualitative and quantitative value, a criterion or a norm. We each have a standard against which we measure how things around us fit into our comfort zone. Our standard is therefore our basis for judging a person, situation, or thing as right or wrong, good or bad, comfortable or uncomfortable. It is not necessarily an accepted norm of social morality, however, because each person's standard is solely their own mental landscape of acceptability and has no validity for anyone else.

One's standard of judgment can be so narrow and biased, however, that it is self-defeating because it blinds them to the truth. Consider the 12-member committee for admissions to a prestigious prep school in New York, which voted unanimously to exclude a particular 13-year-old boy. The rejection was not surprising because the boy's academic record contained marks from failing to barely passing in almost every subject except English. In addition, his teachers' comments about his behavior ranged from "lazy" to "rebellious." After its decision, the committee learned that it had just passed judgment on the scholastic record of young Winston Churchill.[13]

I (Chris) believe everyone—myself included—does the best they know how to do at all times, and I seriously doubt any of us live up to our own standards. If this is true, where is *the* standard, as a true basis for judgment?

> Judgment, like other devices by which the world of illusions is maintained, is totally misunderstood by the world. It is actually confused with wisdom, and substitutes for truth. As the world uses the term, an individual is capable of "good" and "bad" judgment, and his education aims at strengthening the former and minimizing the latter. There is, however, considerable confusion about what these categories mean. What is "good" judgment to one is "bad" judgment to another. Further, even the

same person classified the same action as showing "good" judgment at one time and "bad" judgment at another time. Nor can any consistent criteria for determining what these categories are be really taught. At any time, the student may disagree with what his would-be teacher says about them, and the teacher himself may well be inconsistent in what he believes. "Good" judgment, in these terms, does not mean anything. No more does "bad."[14]

All we can judge is *appearances*. There is nothing else. An appearance is an outward aspect or an outward indication. Judgment is the process of forming an opinion or evaluation by discerning and comparing.

People who are afraid of deviations from their standard cope with their fears by remaining the same while trying to control circumstances so other people will have to risk change—but not them. If the "other" people are unwilling to change, they become enemies, who are judged as not being okay, even inferior, because they do not live up to a certain standard. There are *no enemies out there*, however, only *people frightened by change*, of losing control, of being powerless, and thus they are mistakenly rejected by their fellow human beings.

I (Chris) was a bachelor in the mid-1970s, and everything in my house was just so. Everything had a place and was in its place—always. I was so rigid with my standard of housekeeping that I was often uncomfortable with someone else's.

One day a friend of mine, named Walter, called and said he needed a place to live and asked if I knew of any. Without thinking, I said, "Yes. I have an extra room. Come and live with me." That was the beginning of the end. I did not know that I was about to get an education. He moved in with his dog, horse blankets, bridles, saddles, tools, rifles, and even his periodic girlfriends. My nice, neat, orderly, quiet, simple life was an instant shamble because in those days my friend seemed to be utter chaos looking for a place to unravel.

Yet, this was one of the best things that ever happened to me. I could not "correct" my friend, as it were, no matter how hard I tried, and believe me, I tried. So, I eventually joined him. He, in his mid-20s, taught me, at the age of 40, how to play. He helped me to relax my standards and live a little. He became my personal counselor in overcoming my workaholism and my perfectionism. He gave me an irreplaceable gift: the ability to seize the moment and live it to the fullest. Thank you, Walt.

Victimhood

Feeling like a victim of anything is a helpless, hopeless feeling, a feeling of being somehow violated. Being a victim of abuse is a violent, confusing, and overwhelming experience. One of the major aspects of being a victim is the experience of having little or no control over events. Something happens to you, and you feel powerless to do anything about it. In fact, implicit in the word *victim* is to be at the mercy of events, or of a circumstance or person, essentially to be in a position in which you have no control over what happens to you.

To use being a victim as a coping mechanism, therefore, is difficult at best for a mediator to deal with because it is usually an unconscious act that is readily denied. After all, who would be a victim by choice? Although the definitions that follow seem to be relatively clear-cut, they are intellectual and leave much unsaid.

There are many definitions of victim, but three general ones will suffice for our discussion of victimhood as a coping mechanism: (1) one who is harmed by or made to suffer from an act, circumstance, agency, or condition; (2) a person who accidentally suffers injury, loss, or death as a result of a voluntary undertaking—a victim of their own behavior; and (3) a person who is tricked, swindled, duped, or taken advantage of.

One thing to notice about these definitions is that the victim suffers from a loss of control and is dealt a cruel blow by life. This image of having lost control makes playing a victim an easy way to get out of having to accept responsibility for having made the choice that put you in the circumstance of being the victim in the first place. Most of us probably play the victim at some point in life to cope with the feeling of being humiliated for not possessing control. What our society is telling us, in our own minds at least, is that it is not okay to be human, to err, that loss of control is somehow a terrible weakness.

Nevertheless, to understand and accept the premise of life-affirming mediation requires not only an understanding of coping mechanisms but also an acknowledgment of each person's capacity for rational thought.

The Capacity for Rational Thought

It is critical that we, as mediators, work from the premise that all people possess the potential for rational thought (rational logic). To attain it, however, they must first work their way through the barriers of existing irrational thoughts (emotions that give rise to a sense of logic that is irrational, *especially fear*).

It is thus necessary to understand the meaning and relationship of two words: emotion and logic. As used here, emotion is a state of feeling, such as peace, joy, anger, or fear, which is centered in the individual's sense of self. Emotion is the energy that drives us, that gives us values and feelings. Whereas anger and fear produce an irrational sense of logic, peace and joy are positive emotions and foster rational logic. Rational logic is the mechanism that allows us to understand the emotions contained in our values. Logic allows an individual to understand that they are part of an interconnected, interactive system in which the governing principle of cause and effect is impartial.

Because the inviolate biophysical principles that govern the universe are rational, Nature, which obeys these principles, is rational. We as a part of Nature must therefore possess the potential for rational thought, but we also possess—and often focus on—irrational thought and so imperfectly understand Nature's rationality.

That notwithstanding, unless we believe that the people with whom we work possess the potential for rational thought, we are powerless to bring about peace, short of total human annihilation, because we cannot negotiate with another person as long as we are convinced that our counterpart's thinking is irrational. A cardinal principle of mediation, and the power to act with confidence, is the belief that the people with whom we work possess the potential for rational logic. This means we must believe the people have the potential to honor their feelings while thinking clearly, accept their individual power, and possess the desire for peace. Only then is it possible to reach an accord with them.

When thinking is rational, based on the impartial principle of cause and effect, the group we are working with can become dedicated to the proposition that no person shall abuse another, that all members shall defend the rights of each member, and that each member shall defend the rights of all members. This is possible because a rational person tends to seek peace, which in turn can lead to the organized enactment of a shared vision for the future.

Rational thought can be tested through its converse. Namely, if one does not believe in the rational nature of another person, then one believes it is impossible to negotiate with that person. If one does not believe that rational people ultimately desire peace, then one cannot negotiate confidently toward peace with one's opponent. If one cannot negotiate with one's opponent, one is powerless to achieve peace, and if one cannot organize around rational thought, then the principles of peace cannot move from the minds of people into the actions of society.

However, by the time I (Chris) am normally asked to mediate the resolution of a conflict, the people involved have usually reached a place of such psychological pain, such feelings of fear and rage, that a sense of hopelessness prevails. Under such circumstances, it is often difficult for them to think clearly and therefore direct much attention to dealing with the real causes of their emotions.

Although emotion and logic appear to be mutually exclusive, both are valid because they are different and not substitutable. For example, emotions (either negative, such as pain, fear, and despair, or positive, such as peace, love, compassion, and joy) must be validated and the negative transcended before the impartial logic of the whole systemic picture can be accepted. Thus, negative emotions can be brought to logic only when all parties feel safe enough to be open and honest: where love, trust, and respect can ultimately prevail, when "a gentle answer turns away wrath."[15]

A major problem in the world today is the apparently irreconcilable split between faulty logic based on repressed emotions[16] and rational logic based on the impartial outworking of cause and effect, which is reached when the emotions are accepted, validated, understood, incorporated, and transcended. If not transcended, the faulty logic of unbridled negative emotions can become violence.

Only when negative emotions are transcended can they give the insight, the inner vision, necessary to reach rational thought, that which allows one to perceive the world as an impartial system based on cause and effect. If we, as mediators, choose to accept our own inner struggles toward rational thought, we can help others to do the same just because we have the courage to make that choice.

Everyone Is Right from Their Point of View

We now come to the notion of right versus wrong, again based on individually perceived similarities and differences among one another, as human beings. Therefore, our individual perceptions can be thought of in a manner similar to that of an insect's compound eye because it is through perception that we "see" one another and everything else. The cornea of an insect's compound eye is divided into a number of separate facets, which, depending on the insect, may vary from a few hundred to a few thousand. Each compound eye is formed from a group of separate visual elements, each of which corresponds to a single facet of the cornea. Each facet has what amounts to a single nerve fiber, which sends optical messages to the brain. Seeing with an insect's compound eye would be like seeing with many different eyes at once.

Each human perception is like a facet in the compound eye of an insect, with its independent nerve fiber connecting it to our local, national, and global society (the brain). Thus, each perception, which at the same time represents both an individual's own and his or her own cultural foundation and moral limitations, has its unique construct, which determines the possibilities of the individual's understanding. A person who tends to be negative or pessimistic, for example, sees a glass of water as half empty, while a person who tends to be positive or optimistic sees the same glass of water as half full. Regardless of the way it is perceived, the level of water is the same, which illustrates that we see what we choose to see. And, what we see may have little to do with reality.

So, it seems reasonable that the freer we are as individuals to change our perceptions without social resistance in the form of ridicule or shame, the freer is society—the collective of the individual perceptions—to adapt to change in a healthy, evolutionary way. On the other hand, the more rigidly monitored and controlled "acceptable" perceptions are, the more prone a society is to the cracking of its moral foundation and to the crumbling of its infrastructure because nothing can be held long in abeyance, least of all social evolution.

As there are as many points of view in the compound eye of an insect as there are facets, so there are as many points of view in a society as there are people. Although everyone is right from their point of view, no one person has the complete image, and no one is *totally right*.

We all sense things differently when we see, hear, touch, taste, and smell; because we sense things differently, we understand them differently. In addition, our senses are variously effective under ever-changing circumstances. Our individual brains coordinated and integrated our individual sensing, producing an individual awareness. Through communication, our manifold individual degrees of awareness are coordinated and integrated into a collective awareness. It is through our senses that we become aware of the complementary nature of the "otherness."

It is precisely because we each have our point of view (after we have considered the data and have reached a conclusion, based in large part on the background of our social conditioning) that I (Chris or Lynette) cannot convince you of anything. But, from a point of communication, I must accept what you think you heard, and you must accept what I think I said. Such acceptance is important because for me to convince you that I am right, I must simultaneously convince you that you are wrong. You will resist, of course, first because I have assaulted your dignity and second because you *are* correct from your interpretation of "your" data, just as I am correct from my interpretation of mine and Lynette is correct from hers.

A fascinating learning activity involved 30 of us (Lynette and other participants) standing outside on a park lawn, each of us with one large image that we could not show to anyone, our task was to place them (and ourselves) in order, sequentially, so the imagery told the story these cards represented. We were not to reveal the pictures until we felt we were all in the correct order. The only means to do this was by describing our individual graphics to each other: features, objects, or colors on the pictures that may have represented a dot (miniscule component) or colossal part of the image that another person held. This exercise was based on Istvan Banyai's children's books, *Zoom*[17] and *Re-Zoom*,[18] in which graphic images within images, within images, tell a story through perspective and scale of both temporal and spatial relationships. This fun 30-minute activity calls for the participants to use effective communication skills and processes for finding solutions to challenging issues.

For example, I (Chris) once held a postcard up in front of an audience and asked them to describe it and then tell me what it was. I received a surprising number of different answers. The interesting thing about it was that no one got their description 100 percent correct from a factual point of view—only from their point of view. Moreover, there was some contestation regarding who was correct.

Although I cannot convince you that you are wrong without somehow attacking your dignity, I can give you new data, which *raises the value* of you making a new decision based on new information. In this way, I can be patient and give you the space that allows you to change your mind, if you so choose, while maintaining your dignity intact.

The question, then, is who is right when we are all right from our own points of view? If everyone is right, then who is wrong? Because no one is wrong, we cannot argue any case based on "right" or "wrong." Right or wrong is always a human judgment dealing with appearances, not reality, which means that if I think I am right, I must "win," and if I win, I must be right. You, on the other hand, are clearly wrong because you "lost," and you lost because you are clearly wrong. Thus, each side becomes committed to winning agreement with its point of view and is not even in a position to contemplate another possibility under the competitive illusion of winners and losers.

Everyone loses, however, when issues are "settled" by judgments of right or wrong because everyone appears to be right from their own point of view. This really is no different from a world at war in which each nation, each army, each person is convinced God is on their side.

The duality of right versus wrong does not have to exist. There can instead be a continuum of "rightness" in which some are a little more accurate than others. Such a continuum is predicated on our individual lack of knowledge owing to our own limited perception of possible outcomes. Because we do not know for sure who is righter than whom, we must, in fairness, accept that everyone is right from their own point of view, and each point of view is different—not wrong, only different, regardless of what the discussion is about. The notion of wrong is therefore unacceptable in resolving a conflict.

If we are to survive the present upheavals of social evolution, we must be willing to accept the notion of *right, right, and different*. Wrongness in the classical, combative, human sense must become a relic of the past if we are to treat others as we ourselves would like to be treated. Only then can any issue be *resolved* in such a way that each side retains its dignity and society can progress with some semblance of order into the future.

We find the duality of rightness and wrongness of almost everything to be so pervasive that the notion of right, right, and different is exceedingly difficult to get across in a society that stresses judgmental values as the wisdom of its norm. If we insist on the duality of right versus wrong, we will always be in competition with one another. If, on the other hand, we can agree that everyone is right from their own point of view and that each point of view is only a different perception along the same continuum, we will be able to coordinate and cooperate with one another and together raise the level of our social consciousness to the benefit of all generations.

Acceptance of Circumstances Offers the Choices of What Might Be

Choice equals hope. Choice and hope are the ingredients of human dignity. Dignity means living in peace, free of fear. Our most important choice in overcoming fear and violence is learning to accept a circumstance, whatever it is, as it is. In talking about acceptance, Mother Teresa said (and I, Chris, paraphrase from what I remember) that if God puts us in a palace, to accept being in the palace. By the same token, if God puts us in the street, to accept being in the street. We are not, however, to put ourselves either in a palace or in the street; we are simply to accept whatever circumstance is presented to us at the present moment.

Unconditional acceptance of circumstances is perhaps the most difficult lesson with which I (Chris) struggle daily. I cannot, for example, control a circumstance, but I can choose to control how I respond to it and what my attitude will be. In that choice lies my freedom from fear because I recognize that I have a choice, and I have it *now*. Thus, by giving up trying to control things outside myself, I am in better control *of myself* and can choose how I want to respond to any given circumstance.

In the last analysis, we have a choice. While we always have a choice, *we must choose, in that we have no choice,* much as we might wish this paradox to be otherwise. If you do not like the outcome or if you err in your choice, you can choose to choose again. You are not, therefore, a victim of your circumstances but rather a product of your choices and your decisions based on those choices.

What in a conflict is your choice? Your choice is to control your attitude, as manifested through your behavior. You are thus responsible for your behavior and therein lies the potential resolution of any dispute: to raise the value of changing your attitude and thus your behavior for the common good.

Our task, as mediators, is to help the parties understand that attitudinal and behavioral change is the key to the resolution of their dispute. If one or both of the parties want something, it is up to them to decide how they must behave in order to enhance the possibility of achieving their goals.

The more people are able to choose love and peace over fear and violence, the more they gain in wisdom and the more we all live in harmony. This is true because what we choose to think about determines how we choose to act, and our thoughts and actions set up a self-reinforcing feedback loop, a self-fulfilling prophecy that becomes our reality and our truth.

Choice is the tool with which we, as people, make ourselves who we are. It is here that we must recognize, amidst the myriad choices daily confronting us, that as we think so we create and so we become, either on the material plane or on the spiritual plane. And, we are either freed by our creations (those born of love) or imprisoned by them (those born of fear).

The choice is ours for we have free will in our thoughts and actions, which means that each day, we edit and re-edit our autobiographies. Choice is thus the clay with which we daily mold and remold our character until the day we look into the mirror of our souls and see the cumulative reflection of our many choices, which have manifested as irreversible consequences beyond our control. But, for now, we can still choose what we might be in that future time.

Discussion Questions

1. What is social conditioning?
2. How does our social conditioning cause conflicts?
3. What is dysfunction? How does it affect conflict resolution?
4. What is homeostasis? How is it used? Is it even possible? (See Biophysical Principles 8, 9, 13, and 14 in Chapter 5.)
5. What are coping mechanisms? How do they affect conflict resolution?
6. If everyone is "right" from their point of view, then who is wrong?
7. If everyone is "right" from their point of view, how can a conflict be resolved?
8. Is there a particular question you would like to ask?

Endnotes

1. BrainyQuote. Rabindranath Tagore. http://www.brainyquote.com/quotes/authors/r/rabindranath_tagore.html (accessed February 6, 2010).
2. Buckminster Fuller. http://docs.google.com/viewer?a=v&q=cache:GJJx8PgsnE MJ:www.acceleratedlearning.info/files/Quotes.pdf+Buckminster+Fuller+%22 every+child+is+a+genius&hl=en&gl=us&sig=AHIEtbTyhXT5GdvK69UqbQgD 7V3Vztj9Mg (accessed February 6, 2010).
3. ThinkExist. Hodding Carter. http://thinkexist.com/quotation/there_are_two_lasting_bequests_we_can_give_our/13675.html (accessed February 6, 2010).
4. Edward Bach. *Heal thyself: An Explanation of the Real Cause and Cure of Disease.* Daniel, Essex, England, 1931. 60 pp.
5. AZ Quotes. Mother Teresa. http://www.azquotes.com/quote/392890 (accessed Juned 15, 2017).
6. Robert Frost. Mending Wall. https://www.poetryfoundation.org/poems/44266/mending-wall (accessed July 15, 2018).

7. James W. Tamm and Ronald J. Luyet. *Radical Collaboration: Five Essential Skills to Overcome Defensiveness and Build Successful Relationships*. HarperCollins, New York, 2004, p. 44. 336 pp.
8. Ibid., p. 30.
9. Bach, *Heal Thyself*.
10. M. Durrant. Therapy with Young People Who Have Been the Victims of Sexual Assault. *Family Therapy Case Studies*, 2(1987):57–63.
11. BrainyQuote. Robert Frost Quotes. https://www.brainyquote.com/quotes/quotes/r/robertfros101423.html (accessed June 15, 2017).
12. Leviticus 16:21–22. *The Holy Bible* (Authorized King James Version). World Bible, Iowa Falls, IA.
13. T. J. Fleming and A. Fleming. *Develop Your Child's Creativity*. Association Press, New York, 1970.
14. Foundation for Inner Peace. *A Course in Miracles. Volume 3, Manual for Teachers*. Foundation for Inner Peace, Tiburon, CA, 1975, p. 26.
15. *The Holy Bible*, Proverbs 15:1.
16. (1) Donald Ludwig, Ray Hilborn, and Carl Walters. Uncertainty, Resource Exploitation, and Conservation: Lessons from History. *Science*, 260(1993):17, 36; (2) Chris Maser. *Global Imperative: Harmonizing Culture and Nature*. Stillpoint, Walpole, NH, 1992. 276 pp.
17. Istvan Banyai. *Zoom*. Puffin, New York, 1995. 64 pp.
18. Istvan Banyai. *Re-Zoom*. Puffin, New York, 1995. 64 pp.

7

Communication, the Interpersonal Element

Ideally, mediation revolves around understanding and sharing emotions and knowledge, both of which grow from and are a reflection of social experience. Emotions and knowledge are shared through communication, which is the very heart of conflict resolution and must be treated with the utmost respect. Just as dishonest or careless communication tells much about the people we are listening to, so does good communication. Good communication means respect for both listener and speaker because one must first listen to understand and then speak to be understood.

Communication is perhaps one of the most difficult things we do as human beings, yet it is simultaneously one of the most important things. We are creatures who must share feelings, senses, abstractions, and concrete experiences in order to know and value our existence in relation with one another. Communication, or language, is the way we share the very core of our relationships. Our existence revolves around it. Without it, we have nothing of value.

Language as a Tool

Although communication involves far more than mere words, we are concerned only with words here. Words are symbols for the things we experience; therefore, the more accurately a chosen word builds a bridge to our common understanding, the easier it is to get in touch with one another, stay in touch, build trust, and ask for and receive help.

In this sense, semantics is more than quibbling over words; it reveals both our thought patterns and our consciousness of cause and effect. It is the conveyance of concepts, perceptions, personal truths, trust, and a shared vision for the future. Like every linguistic creation, language can empower or limit, depending on whether we see it as a set of labels describing some preexisting, unchangeable reality or as a medium with which to articulate a new reality, a sustainable way of living together on and with Earth.

Our task as a mediator is to create a safe place in which common bonds can be built, maintained, and strengthened through good communication or a clear, concise use of language. Just as any relationship requires sensitive, honest, and open communication to be healthy and grow, so are relationships in the mediation process forged, maintained, and improved when feelings and information are shared accurately, freely, and with sensitivity.

The quality of communication is thus enhanced if simple, rather than complex, words are used. Picturesque slang and free-and-easy colloquialisms, if they are appropriate to the subject and if they do not offend the sensibilities of the participants, can add variety and vividness to the mediation process. But, substandard English, such as grammatical errors and vulgarisms, not only detracts from the mediator's dignity but also reflects the mediator's attitude toward the participants' intelligence.

If the subject under discussion includes technical terms, the mediator must be sure to define each term clearly and concisely so that all participants know exactly what is meant by it. It is also best to use specific, rather than general, words. In addition, to ensure clarity, short- to medium-length sentences are best because, for most people, the spoken word is often more difficult to grasp than the written word, which not only is visible but also can be read over and over and studied. The words used and how we define them are extremely important.

Quality communication requires consistent practice. It is thus imperative that Lynette and I continually monitor the words we use, their meanings, and their usage in our everyday speaking and writing. Every word must be valued, and any word that does not carry its weight must be discarded. We (Lynette and I) must practice all day, every day, because good communication comes first and foremost from good thoughts. By our thoughts, we (in the generic) privately define and by our actions we publicly declare who and what we are. British psychologist James Allen stated this beautifully:

> A man's [and a woman's] mind may be likened to a garden, which may be intelligently cultivated or allowed to run wild; but whether cultivated or neglected, it must, and will, bring forth. If no useful seeds are put into it, then an abundance of useless weed seeds will fall therein, and will continue to produce their kind.
>
> Just as a gardener cultivates his plot, keeping it free from weeds, and growing the flowers and fruits which he requires, so may a man tend the garden of his mind, weeding out all the wrong, useless, and impure thoughts, and cultivating toward perfection the flowers and fruits of right, useful, and pure thoughts. By pursuing this process, a man sooner or later discovers that he is the master gardener of his soul, the director of his life. He also reveals, within himself, the laws of thought, and understanding, with ever-increasing accuracy, how the thought forces and mind elements operate in the shaping of his character, circumstances, and destiny.[1]

A word spoken is thus the manifestation of a thought, whether positive or negative. Once spoken, it can never be withdrawn, despite an apology, because words are the public extensions of our private selves. Communication, whether spoken or written, has the power to build or corrode trust.

The "water message" game (which can be modified for any shared natural resource situation) is a tool that helps demonstrate the importance and delicacy of trust and shows that, once broken, it cannot readily be repaired.[2] This hour-long exercise that I (Lynette) use in the classroom also illustrates the degrees to which win-loss outcomes impact the ability of forming water alliances. The premise is centered on a communal reservoir; parties work separately for their individual communities toward economic growth (new technologies, industries, and agriculture) by means of water. Between the parties and through a neutral facilitator, only three types of messages are exchanged on a slip of paper: (1) "We maximize our water use for the greatest economic growth," which communicates an attitude of not caring about the water needs of others; (2) "Our water use is balanced, which affords us measured economic growth," which implies that there is a shared approach to water use that considers the needs of others; or (3) "Our water use is minimal, so our economic growth is stifled," which suggests a hands-off neutral approach regarding the water use for anyone else.[3] Highest scores are earned per round, when cordial exchanges occur between parties and there are sustainable economic gains; this only occurs when trust is high. So, within the game, the key is finding and maintaining balance between resource needs and building trust (through favorable relations and good communication). Good communication, a prerequisite for both teaching and learning, clears the way for shared, participative ownership of ideas as a means of building relationships within the mediation process. There are, however, a number of obligations that accompany good communication.

Because the right—and need—to know is basic in the mediation process, all parties must have equal access to pertinent information if a conflict is to be resolved. Here, we believe, it is better to err on the side of sharing too much information rather than risking someone being left in the dark. We say this because hoarded information subverts the mediation process through misrepresentation.

Everyone has a right to simplicity and clarity in communication and an obligation to communicate simply and clearly, especially a mediator. If I (Chris) cannot be simple and clear in what I say, then I do not understand the topic well enough to discuss it. Sometimes, for example, I have trouble expressing a concept while writing a book, giving a speech, or during the facilitation process. When this happens, I write an essay on the topic in a maximum of five double-spaced pages. And, I work on it until I have got it down in the best manner I can. From that point, it is clearly in mind and flows easily whenever I need to discuss the subject.

I go to this length because I owe everyone truth and courtesy, although truth is often uncomfortable and at times a real constraint, and courtesy may be an inconvenience. Nevertheless, it is these qualities that allow communication to educate and liberate us.

I am obliged to practice discrimination in both what I say and what I hear, which means that I must respect my own language through its careful usage.

I must acknowledge and accept that muddy language means muddy thinking, and muddy thinking means muddy communication. I must therefore always remember that my audience (the parties in the dispute) may need something special from me, such as an extraordinary amount of patience and clarity, while they struggle to communicate.

Language is the most profound tool Lynette and I have as mediators because it both educates and liberates. Teaching and learning underlie mediation literacy and action. Mediation literacy is the "why" the process does what it does, and action is the "what" it does. With this in mind, we can use language to help the parties engaged in a mediation process free themselves from their shackles of conflict. To allow the parties to liberate themselves, however, our communication must be based on sound reasoning, compassion, detachment, and sometimes on silence.

The Use of Silence in Communication

A mediator must learn to appreciate the power of silence in communication. Most people are profoundly uncomfortable with silence and feel compelled to speak, including many mediators who possess the need to direct, control, and intervene in the process—and so destroy it. Silence, when allowed to flow unimpeded through indeterminate seconds and minutes, draws people out, causing them to engage both uncomfortable circumstances and one another.

For example, some years ago a small group of ranchers in central Oregon asked me (Chris) to help them articulate a vision statement for grazing livestock on public lands, one that would allow them to continue using public lands if they lived up to it. I agreed, but said that I could only help them if they could tell my why they wanted to be ranchers. Silence. They had no answer. They had never thought about it. So, I took out the book I was reading, sat down, and became engrossed in it.

After agonizing over the question for a couple of hours, one of the ranchers approached me and said, "Ahh . . . , Chris, ahh . . . , I guess it's the way of life that I love."

"Okay," I replied, "let's call it lifestyle. What's it worth to you? How much are you willing to change your attitude and behavior concerning your use of public lands to maintain your chosen lifestyle?"

With these questions answered, the ranchers drafted their vision statement. Now, years later, they not only are grazing their livestock on public lands but also are model ranchers. One of them traveled around the United States speaking to other ranchers about what he had learned.

It was their agony in the 2-hour silence that finally drew out the answer they needed to find, and the answer was theirs—not mine. Had I in any way helped them with the answer because I was uncomfortable with the silence, it would have been my answer—not theirs—and it would have been useless to them. As it turned out, the answer raised the value of being ranchers. They felt like the first-class citizens they were, and because they acted accordingly, people listened when they spoke.

The Need to Be Heard

Although one may not think of it as such, listening is the other half of communication. Communication is a gift of ideas; therefore, the other person can give you a gift of ideas through speaking only if you accept the gift through listening. The spoken word that falls on consciously "deaf ears" is like a drop of rain evaporating before it reaches Earth. Intolerance of another's ideas belies faith in one's commitment to resolving the conflict.

The watchword of listening is *empathy*, which means imaginative identification with, as opposed to judgment of, the person's thoughts, feelings, life situation, and so on. The more a mediator can empathize with a person, the more that person feels heard, the greater the bond of trust, and the better you (as mediator) can understand the situation. This means, however, actively, consciously listening with a quiet, open mind, without forming a rebuttal while the other person is speaking. Such listening is an act of love, and anything short of it is an act of passive violence.

I (Chris) was once on a television program in which the intent was to discuss the issue of ancient forests in the Pacific Northwest. An elderly lady on the program tried in vain to be heard, but the moderator consistently ignored her. Even after we were off the air, she tried again to tell the moderator how she was feeling, but he continued to ignore her. In the end, just to be heard, perhaps only by herself, she spoke out loud to no one; she spoke into space. She may as well have been alone in the world.

Not listening is an act of violence because it is a purposeful way of invalidating the feelings—the very existence—of another person. Everyone needs to be heard and validated as a human being because sharing is the essence of relationship that makes us "real" to ourselves and gives us meaning in the greater context of the universe. We simply cannot find social meaning outside a relationship with one another. Therefore, only when we (in the generic) have first validated another person through listening, as an act of respect, can that person really *hear* what *we* are saying. Only then can we share another's truth. Only then can our gift of ideas touch receptive ears.

All we have in the world, as human beings, is one another, and all we have to give one another is one another. We are each our own gift to one another and to the world; we have nothing else of value to give. We cannot give our gift, however, if there is no one to receive it, there is no one to hear. Therefore, if we listen—really listen—to one another and validate one another's feelings, even if we do not agree, we can begin to resolve our differences before they become disputes. But, to listen well and to speak well, it is important to consider the basic elements of communication.

The Basic Elements of Communication

Communication occurs when one person transmits ideas or feelings to another or to a group of people. Its effectiveness depends on the similarity between the information transmitted and that received—a common frame of reference.

The communication process is composed of at least three elements: (1) the sender, who is someone speaking, writing, signing, or emitting the silent language of attitude or movement; (2) the symbols used in creating and transmitting the message, which can be sounds of a particular and repetitive form called spoken words, particular and repetitive handcrafted signs called written words, a particular arrangement of musical notes called melody, and facial expressions, hand motions for the deaf, touch for the blind, and generalized "body language" for the sighted; and (3) the receiver, who is someone listening to, reading, observing, or silently feeling the symbols through touch. These elements are dynamically interrelated, and that which affects one influences all.

Suppose Lynette has something she wants to convey to you in her teaching about water conflict resolution. She sends her thoughts through the air, as intelligent noise, for you to pick up with your receivers, your ears. You must then translate the sounds back into your thoughts, as you understand them. And, you think you know what she did her best to convey? She cannot accurately tell you what she meant because there are no words to express the nuances of thought. How, for example, can she really say, "I love you," to her child? What does that mean? She can feel it, but there simply are *no words to transfer the feeling* to her child because words are only metaphors we use to talk around the sense of feeling or meaning we cannot directly convey.

Communication is thus a complicated, two-way process that is not only dynamic among its elements but also reciprocal. If, for instance, a receiver has difficulty understanding the symbols and indicates confusion, the sender may become uncertain and timid, losing confidence in being able to convey ideas. The effectiveness of the communication is thus diminished. On the other hand, when a receiver reacts positively, a sender is encouraged and adds strength and confidence to the message. Let us examine how the three elements work.

Sender

A sender's effectiveness in communication is related to at least three factors. First is facility in using language, which influences the ability to select those symbols that are graphic and meaningful to the receiver.

Second, senders, both consciously and unconsciously, reveal their attitudes toward themselves, toward the ideas they are transmitting, and toward the receivers. These attitudes must be positive if the communication is to be effective. Senders must indicate that they believe their message is important and thus necessary to know the ideas presented.

Third, a successful sender draws on a broad background of personal, accurate, up-to-date, stimulating, and relevant information. A sender must make certain that the ideas and feelings being transmitted are relevant to the receiver. The symbols used must be simple, direct, and to the point. Too often, however, a sender uses imprecise language or technical jargon that is nonsense to the receiver and thus impedes effective communication.

Symbols

The most basic level of communication is achieved through simple oral and visual codes. The letters of the alphabet, both spoken and written, constitute such a basic code when translated into words, as do common gestures and facial expressions. But, words and gestures only communicate ideas when combined in meaningful wholes: speeches, sign language, sentences, paragraphs, or chapters. Each part is critical to the meaning of the whole.

Ideas must be carefully selected if they are to convey messages that receivers can understand and to which they can *respond*. Ideas must be analyzed to determine which are best suited for starting, carrying, and concluding the communication and which clarify, emphasize, define, limit, or explain the context—all of which form the basis of effective transmission of ideas from the sender to the receiver.

Finally, the development of ideas from simple symbols culminates in the selection of the medium (such as hearing, seeing, touch, or some combination of the three) best suited for their transmission according to the receiver's abilities. In the mediation process, however, a variety of media (hearing, seeing, touching, and at times smelling and tasting) makes for the most effective communication because it relates to the widest range of experiences.

Receiver

A basic rule of mediation, with respect to our responsibility as mediators, is to be clear, concise, and relevant, and because communication is a shared responsibility, the receiver must do their best to understand. We know communication has occurred when receivers respond with an understanding that allows them to change their behavior.

To understand the communication process, it helps to appreciate at least three aspects of receivers: their abilities, attitudes, and experiences, which often, and in many hidden ways, relate to their familial upbringings. First, it is important to discern a receiver's ability to question and comprehend the ideas transmitted. We can encourage a receiver's ability to question and comprehend by providing a safe atmosphere that welcomes such participation.

Second, a receiver's attitude may be one of resistance, willingness, or passivity. Whatever the attitude, we must gain the receiver's attention and retain it. The more varied, interesting, and relevant we are as mediators, the more successful we will be in this respect.

Third, a receiver's background, experience, and education (often extremely diverse in a group situation) constitute the frame of reference toward which the communication must be aimed. Lynette and I assume the obligation of assessing the receiver's knowledge and of using it as the fundamental guide for effective communication. For us to get a receiver's response, however, we must first reach them, something a growing loss of words makes increasingly difficult with younger generations.

Changes in the *Oxford Dictionary for Children*

Every human language—the master tool representing its own culture—has its unique construct that determines both its limitations and its possibilities in expressing myth, emotion, ideas, and logic. One of the greatest feats of humanity is the evolution of written language: those silent, ritualistic marks with their encoded meaning that not only made culture possible but also archive its history as part of its cultural commons.

The relative independence with which cultures evolve creates their uniqueness, both within themselves and within the reciprocity they experience with one another and their immediate environments. Each culture, and each community within that culture, affects its environment in a specific way and is accordingly affected by the environment in a particular way. So it is that distinct cultures in their living create culturally designed landscapes, which in some measure are reflected in the myths they hold and the languages they speak. As such, language is the medium with which the condition of the human soul is painted.

The artist, using words to convey the colors of meaning by mixing them on a palette of syntax, composes the broad shapes of a cultural story line. Then, by matching the colors of words to give expression to ideas, the artist adds verbal structure, texture, and shades of meaning to the story. In doing so, the verbal artist paints a picture or portrait as fine as any accomplished with brush, paint, palette, and canvas; with camera and film; or with musical instruments and mute notes on paper. In addition, a verbal picture often outlasts the ravages of time, which claim those of paint on canvas, imprints of light on photographic paper, or recordings of musical instruments that give "voice" to mute shapes.

So, what does it say about Western industrialized society when the latest edition of the *Oxford Junior Dictionary* has omitted words of historical significance pertaining to Nature and culture to make way for greater modernity, including such "technobabble" as *BlackBerry, blog, voice mail,* and *broadband*?

Yet, according to Vineeta Gupta, head of the children's dictionaries at Oxford University Press, changes in the world are responsible for these alterations, "When you look back at older versions of dictionaries, there were lots of examples of flowers for instance. That was because many children lived in semi-rural environments and saw the seasons. Nowadays, the environment has changed." Several criteria were used to select the 10,000 words and phrases in the junior dictionary, including how often words would be used by young children.[4] However, as Elaine Brooks points out, "Humans seldom value what they cannot name."[5]

Nature words deleted from the Oxford Junior Dictionary include acorn, adder, almond, apricot, ash, ass, beaver, beech, beetroot, blackberry, bloom, bluebell, boar, bramble, bran, bray, brook, budgerigar, bullock, buttercup, canary, canter, carnation, catkin, cauliflower, cheetah, chestnut, clover, colt, conker, corgi, cowslip, crocus, cygnet, dandelion, doe, drake, fern, ferret, fungus, gerbil, goldfish, gooseberry, gorse, guinea pig, hamster, hazel, hazelnut, heather, heron, herring, holly, horse chestnut, ivy, kingfisher, lark, lavender, leek, leopard, liquorice, lobster, magpie, melon, minnow, mint, mistletoe, mussel, nectar, nectarine, newt, oats, otter, ox, oyster, panther, pansy, parsnip, pasture, pelican, piglet, plaice, poodle, poppy, porcupine, porpoise, poultry, primrose, prune, radish, raven, rhubarb, spaniel, spinach, starling, stoat, stork, sycamore, terrapin, thrush, tulip, turnip, vine, violet, walnut, weasel, willow, wren.

Cultural words taken out of the dictionary: abbey, aisle, allotment, altar, bacon, bishop, blacksmith, bridle, chapel, christen, coronation, county, cracker, decade, devil, diesel, disciple, duchess, duke, dwarf, elf, emperor, empire, goblin, manger, marzipan, monarch, minister, monastery, monk, nun, nunnery, parish, pew, porridge, psalm, pulpit, saint, sheaf, sin, vicar.

Words put in: allergic, alliteration, analogue, apparatus, attachment, bilingual, biodegradable, block graph, blog, boisterous, brainy, broadband, bullet point, bungee jumping, cautionary tale, celebrity, chatroom, childhood, chronological, citizenship, classify, colloquial, committee, common sense, compulsory, conflict, cope, creep, curriculum, cut and paste, database, debate, democratic, donate, drought, dyslexic, emotion, endangered, EU, Euro, export, food chain, idiom, incisor, interdependent, MP3 player, negotiate, square number, tolerant, trapezium, vandalism, voicemail.[6]

Here, the challenge is that ideas, which depend on words for conveyance, breed awareness $\Rightarrow$ understanding $\Rightarrow$ consciousness $\Rightarrow$ choices $\Rightarrow$ initial adaptability $\Rightarrow$ decisions $\Rightarrow$ actions $\Rightarrow$ trade-offs $\Rightarrow$ irreversible consequences ⇒ and so on. Therefore, when words are omitted from the cultural lexicon, the art of language is diminished, as is the ability to understand the biophysical systems that support the growing technological isolation from one's natural environment.

Some languages, as exemplified previously, are simply being eroded through the conscious substitutions of words, whereas others cease to exist altogether. Although language is not something we generally think of as becoming extinct, languages are disappearing all over the world, especially the spoken-only languages of indigenous peoples. As languages vanish, so do the cultural variations of the landscape they allowed, even fostered, because a unique culture cannot exist without the uniqueness of its language to protect its history and guide its evolution.

While it probably took thousands of years for the different human languages to evolve, it can take less than a century for some of them to disappear. As languages become extinct, we lose their cultural knowledge along with their perceptions and modes of expression. Because language is the fabric of culture and the living legacy of our identity, when a language dies, the demise of the culture that gave it birth is imminent.

What is lost when a language becomes extinct? How many potential answers to contemporary problems and how much ancient wisdom will be lost because we are losing languages, especially obscure, indigenous ones, to "progress"?

With the loss of each language, we also lose the evolution of its logic and its cultural myths and rituals, those metaphors that give the people a sense of place within the greater context of the universe, because language represents unity within and through time. Temporal unity is the language of memory, those images of experience stored in the human psyche and passed forward from generation to generation in the form of stories, myths, and rituals. Therefore, each time we allow a human language to become extinct, we are losing a facet of understanding, a facet of ourselves—the collective memory of a people archived in their language, a memory that is part of the human hologram, our collective commons of the human experience. As a global society, we are slowly making ourselves blind to our relationships with one another, the universe, and ourselves, which is augmented by "Nature deficit disorder" in today's children.

Nature Deficit Disorder in Children

Everyone will likely agree that children today spend most of their time "plugged in" to a game system, a computer, or other electronic device. According to Richard Louv (author of *Last Child in the Woods: Saving Our Children from Nature-Deficit Disorder*), "Nature deficit disorder describes the human costs of alienation from Nature, among them: diminished use of the senses, attention difficulties, and higher rates of physical and emotional illnesses."[7] Children today spend hours "plugged in" to various electronic devices. The causes of *attention deficit hyperactivity disorder* (ADHD) are not all known yet, but recent studies have shown that for each hour of television a preschooler watches per day, their risk of developing concentration problems and other symptoms of ADHD by age 7 increases by 10 percent.[8] Even when they do spend time in "Nature," it is usually not free-play time, but scheduled or structured time, such as at soccer games or other sports.

Parents contribute to this shunning of the outdoors by instilling fear of the outdoors in their children. How many children are allowed to walk to school or play at a nearby park alone? Do we let them get dirty and truly play or complain about them ruining their clothes? Do we encourage free play in the natural environment or buy the SUV with dual video players in it? And, the bigger question, how are we, as environmental trustees, to communicate the value of our natural resources to a new generation that has not experienced Nature?

The bottom line is the less time our children (and our peers) spend connecting with and learning to appreciate the natural world, the more difficult our jobs as environmental trustees, conflict mediators, and human beings will become. We no longer have a society with a shared experience of Nature. As a result, our jobs as environmental conflict mediators have become much more complex and challenging.

Barriers to Communication

The nature of language and the way in which it is used often lead to misunderstandings and conflict. These misunderstandings stem primarily from three barriers to effective communication: (1) the lack of a common experience or frame of reference, (2) how one approaches life, and (3) the use of abstractions.

Lack of a Common Experience or Frame of Reference

The lack of a common experience or frame of reference is probably the greatest barrier to effective communication. Although many people believe that words carry meaning in much the same way as a person transports an armful of wood or a pail of water from one place to another, words *never* carry precisely the same meaning from the mind of the sender to that of the receiver. Words are vehicles of perceptive meaning. They may or may not supply emotional meaning as well. The nature of the response is determined by the receiver's past experiences surrounding the word and the feelings it evokes.

Feelings grant a word its meaning, which is in the receiver's mind and not in the word itself. Because a common frame of reference is basic to communication, words in and of themselves are meaningless. Meaning is engendered when words are somehow linked to one or more shared experiences between the sender and the receiver, albeit the experiences may be interpreted differently. Words are thus merely symbolic representations that correspond to anything to which the symbol is applied by people: objects, experiences, or feelings.

Thus, a sender must differentiate carefully between the symbols and the things they represent, keeping both in as true a perspective as possible. The truth of a perspective (the interpretation of an experience) is based on the degree to which a person is functional or dysfunctional, which is largely determined by the functionality of one's family of origin and one's subsequent social conditioning. It is also based not only on the degree to which a person has grown beyond their own dysfunction but also in the breadth and depth of their individual life experiences. Taken together, this translates into generalized personality traits.

Generalized Personality Traits

In a sense, generalized personality traits are an amalgamation of the dominant coping mechanisms with which one navigates life. They thus become the essence of one's interpretation of life experiences and the springboard of one's personal capabilities. These traits, which we each possess to a greater or lesser degree, are not cut and dried but rather are overlapping tendencies with varying shades of gray. Nevertheless, they can be substantial barriers to communication.

For example, some people can take ideas seemingly at *random* from any part of a thought system and integrate them; these people have mental processes that instantly change direction, arriving at the desired destination in a nonlinear, intuitive fashion. Others can think only in a *linear sequence*, like the cars of a train; these people have mental processes that crawl along in a plodding fashion, exploring this avenue and that, without assurance of ever reaching a definite conclusion. If the random thinker is also at ease with *abstractions* but the linear sequence thinker requires *concrete* examples, their attempts to communicate may well be like two ships passing in a dense fog.

Then, there is the *introverted* person, who (appearing self-possessed, even aloof) processes things internally, navigates life's path more or less alone, and has few friends over a lifetime. An *extroverted* person, on the other hand, is outgoing, mingles easily with other people, requires the presence of people to be happy, processes things through mutual discussion, and has a constant string of friends. An introvert works well alone behind the scenes, whereas an extrovert works well with people out front. In addition, there are four other types of individuals, which can be summed up as fatalist, exasperator, appraiser, and relator.

A *fatalist* is the consummate victim who feels powerless in the face of an all-powerful system or life itself and is forever suffering a loss of control. To this person, the operational word is *can't*. A fatalist, resigned to their lot in life, is often barely functional and requires a tremendous amount of energy to even reach zero on the scale of enthusiasm—energy sucked from whomever will give it. Just as soon as the person stops propping up the fatalist, however, the fatalist plunges below zero again.

Although the fatalist wants to be rescued, they resist attempts to be rescued at all cost. Here, we mediators must be wary. The only one who can rescue a person is the person in need of rescuing. And, only they know when they are ready for self-rescuing.

Nevertheless, fatalists are good technicians. They tend to be most comfortable with simple, clear instructions about which they do not have to think. Having said this, it is critical to understand that fatalists are usually paralyzed by having to bear responsibility. They work well behind the scenes, are usually patient with details, and may even accept monitoring the progress of an activity, provided they do not have to accept any responsibility for its outcome.

An *exasperator*, on the other hand, must be the center of attention and is deeply invested in so being. Here, the watchword is *control*. Some exasperators go to great lengths to command attention and be in control of whatever involves them. They tend, for instance, to be good at "one-liners," know "all" the jokes, are often the life of the party, and will argue any and every side of an issue, even changing sides in midstream, rather than acquiesce.

Individuals with exasperator personalities are as persistent as bulldogs. Rather than agree, they will say, "Yes, but . . . ," just as long as someone will try to show them another way of thinking about something or another possible outcome.

I (Chris) find it best to openly and freely acknowledge the exasperator's point of view, the supposed position of power, which does not mean that I necessarily agree with it. Once exasperators feel they have exerted their power and have been appropriately recognized, they can relax, and everyone can get on with the process of resolving the conflict.

Once an exasperator has an idea in mind, however, they become impatient for action and, throwing caution to the wind, often barge ahead without getting adequate data or listening to other sides of an argument. On the flip side, if something needs to be done, done well, and completed on time, give it to an exasperator because they will move Heaven and Earth to show off their prowess.

While an exasperator often "knows it all," an *appraiser* wants facts, facts, and more facts. An appraiser seems to be uncertain in the world and wants to make sure that all the data are in, examined in minute detail, weighed, reexamined, and reweighed before any decision is made. Such caution demands much patience on the part of a mediator because an appraiser often seems to hold the forward motion of the process in abeyance, regardless of how much data are at hand. If data are needed, ask an appraiser to obtain it, and you will likely get the best there is—and lots of it.

Then, there is the *relator*, the person who is vitally concerned with what others will think and will go wherever the political wind blows. The relator seldom seems to know who they are as an individual and seems to have ideas only in relation to their acceptability to others. Such a person changes their mind often and gives away their power to whoever asks for it.

Because success or failure is not an event but rather the interpretation of an event, successes or failures of relators are determined by what everyone else thinks because relators are constantly comparing themselves to those around them and internalizing what they are told by others. Unfortunately, we (in the generic) usually lose in the end when we compare ourselves to others because we tend to select someone we admire and then find our differences to be deficiencies, even liabilities.

Relators, in my (Chris's) experience, are subject to getting their feelings hurt easily and often. This is perhaps the major way in which they try to control uncomfortable circumstances because it causes most people around them to "walk on eggshells."

In working with relators, it best to refuse to accept their power, even when it is offered. Instead, ask them what they think and how they feel in an effort to draw them out. Done gently and patiently, this can work quite well.

Relators are generally excellent with public relations because they are sensitive to how others feel and work very hard to win approval. Therefore, they have a good sense of how to market an idea.

There are also product-oriented thinkers, symptomatic thinkers, and systemic thinkers. A *product-oriented thinker* is a person oriented to seeing only the economically desirable pieces of a system and seldom accepts that removing a perceived desirable or undesirable piece can or will negatively affect the productive capacity of the system as whole. This person's response typically is, "Show me; I'll believe it when I see it." To such a person, mediation is usually seen as an immediate problem-solving exercise.

In my (Chris's) experience as a mediator, the more a person is a product-oriented thinker, the more reticent the person is to change. This type of individual sees change as a condition to be avoided because they feel a greater sense of security in the known elements of the status quo, especially when money is involved. But, as Helen Keller once said, "Security is mostly a superstition. It does not exist in Nature. . . Life is either a daring adventure or nothing."[9] (Conversely, the more of a *systems thinker* a person is, the more likely the person is to agree with Helen Keller and risk change on the strength of its unseen possibilities.)

A product-oriented thinker is likely to be a rural resident who is very much concerned with land ownership and property rights and wants as much free rein as possible to do as they please on their own property, at times without regard for the consequences for future generations. The more a person is a product-oriented thinker, the greater the tendency is to place primacy on people of one's own race, creed, or religion, as well as on one's own personal needs, however they are perceived. The more of a product-oriented thinker a person is, the greater the tendency is to disregard other races, creeds, or religions, as well as nonhumans and the sustainable capacity of the land. Also, the more of a product-oriented thinker a person is, the more black and white one's thinking tends to be, as illustrated in the following example:

> The wimpy [sic] comments by Mike Mitchel in Sunday's "Rural Issues" were disturbing. As the head "honcho" and decision-maker for a BLM [Bureau of Land Management] office, he said things like, "We just follow the regulations and enforce them. . . . " Also, "We have our regulations and have no choice."
>
> That's typical bureaucratic arrogance, and a cop-out. Those regulations didn't come down the mountain on stone tablets. They are the product of a well-funded lobby in Washington, D.C., that represents those who are "saving us" from the horrible ranchers, miners and farmers of Nevada.
>
> He says the land will restore itself in 15 or 20 years if we change grazing practices. Restore itself for what? So, some manicured marshmallow-butt from Washington can start up a cattle ranch on abandoned land? Get real! Nevada ranchers are on the land now! The BLM should help them do what they do best, or get the hell out of the way. I'm not to [sic] smart, but I recognize typical Sierra Club rhetoric when I hear it.
>
> As the song goes: When will they ever learn?[10]

As indicated in the Introduction to this book, a *symptomatic thinker,* by analogy, is one who goes to their doctor because they do not feel well. In turn, their doctor tells them that they must get more exercise and lose 20 pounds. To which they respond, "Can't you just prescribe a pill? I don't want to change my lifestyle." If, on the other hand, we modern humans are to live with any measure of dignity and comfort, the symptomatic rationale embedded in our contemporary lifestyles must give way to a systemic approach that recognizes and accepts the reciprocal interactions among all aspects of social-environmental sustainability worldwide. Granted, this sounds like a daunting task, yet in our view it is paramount to human survival. Our hope is to encourage the acceptance of this challenge and to make the case that a more life-enhancing path exists than the dismal course we are now following toward a predictable culmination of great suffering and widespread deprivation.

In contrast to both a person who is a product-oriented thinker or a symptom-oriented thinker, *systemic thinkers* tend more toward a systems approach to thinking. A systems thinker sees the whole in each piece and is therefore concerned about tinkering willy-nilly with the pieces because they know such tinkering might inadvertently upset the desirable function of the system as a whole. A systems thinker is also likely to see him- or herself as an inseparable part of the system instead of apart from and above the system. A systems thinker is willing to focus on transcending the issue in whatever way is necessary to frame a vision for the good of the future.

A systems thinker is not only a rarity but also most often an urban dweller who is likely to be concerned about the welfare of others, including those of the future and their nonhuman counterparts. Systems thinkers also tend to be concerned with the health and welfare of Planet Earth in the present for the future. And, they readily accept shades of gray in their thinking.

There are still other generalizations that can be made, such as people who are visually oriented as opposed to those who respond to sound or touch. In addition, these traits come in a variety of combinations, which indicates how different and complex people can be in response to their life experiences. These differences and complexities naturally carry over into people's patterns of communication. None of these patterns is better than any other as far as the mediation process is concerned; each is only different and needs to be understood.

In the end, we (in the generic), having incorporated all of our familial pieces within ourselves in one way or another (both functional and dysfunctional), go out into the world and take our families, and consequently social conditioning, with us. How we grow up thus determines how we approach life.

As I (Chris) finally broke the cycle of dysfunction within myself, I learned something that is critical to resolving conflicts of any kind: The more a person is drawn toward a peaceful, optimistic view of the future and life in general, the more functional (psychologically healthy) the individual is. Conversely, the more a person is drawn toward debilitating, destructive conflict; cynicism; and pessimism about the future and life in general, the more dysfunctional (psychologically unhealthy) the person is.

Making Language Real

Recognizing and understanding the nuances of how people use language provide vital clues to one's personality dynamics. Such understanding is critical in altering the dynamics of a conflict and helping disputing parties to resolve their differences with dignity.

Concrete words refer to objects a person can directly experience. Abstract words, on the other hand, represent ideas that cannot be experienced directly. They are shorthand symbols used to sum up vast areas of experience or concepts that reach into the trackless time of the future. Albeit they are convenient and useful, abstractions can lead to misunderstandings.

The danger of using abstractions is that they may evoke an amorphous generality in the receiver's mind and not the specific item of experience the sender intended. The receiver has no way of knowing what experiences the sender intends the transmitted abstraction to include. For example, it is common practice to use such abstract terms as "proper method" or "shorter than," but these terms alone fail to convey the sender's intent. What exactly is the "proper method?" Shorter than *what*?

When abstractions are used in mediation, they must be linked to specific experiences through examples, analogies, and illustrations. It is even better to use simple, concrete words with specific meanings as much as possible. In this way, the mediator gains greater elucidation of the images produced in the receiver's mind, and language becomes a more effective tool.

Since I (Chris) mediate the resolution of environmental conflicts, I endeavor to get the participants physically out of the comfortable conference room and into the field, where we can wander through the area of conflict and discuss it. I can thus transform the abstractions of the conference room into concrete examples of the field, which one can see, touch, smell, hear, and, if necessary, taste.

For example, I was asked to mediate a better understanding between a local mill owner/logger and the personnel of the US Forest Service in the state of Colorado. The mill owner had recently purchased a large, very expensive piece of logging equipment that was more efficient in harvesting trees than were the men that used to work for him, who the machine had replaced. The problem was that the huge piece of equipment was severely compacting the fragile soils of the mountainous forest, to the ecological detriment of the sustainable, productive future of the forest. But, the mill owner did not understand the ecological effects he was causing.

I spent 2 days with the Forest Service folks and the mill owner. The first day was spent in the conference room, where I showed slides of how forest ecologists, at that time, thought a forest functioned both above- and belowground. The audience was asked to explore the consequences of long-term management decisions on both the native forest ecosystems and their human culture. Much of the first day's discussion was a maze of abstractions to the mill owner, no matter how simply I explained the data, because he had no frame of reference for what happened belowground.

The second day was spent in the forest in the area where the logging was taking place. We discussed the concepts and data from the first day as we examined and discussed the actual soil condition in the uncut forest and compared it to the area just logged. Throughout the discussion, which included digging in and examining the soil to establish a common frame of reference, I related our observations to the abstract notions of the day before.

By midafternoon, yesterday's abstractions became today's concrete examples to the mill owner, and he began to understand what the Forest Service folks had been trying to tell him. Finally, much to my surprise, he turned to the forest supervisor and said that he had never really understood the consequences of his actions on the forest and on his young son's options to log, if he so chose. He said that while the piece of equipment was more efficient and cost effective than the men who used to work for him, it was compacting the soil so much that the forest may have trouble coming back. He said he realized that by making a little more money now, he might be costing his son the opportunity to log in the future, and he owed his son that chance. Thus, he thought that he should sell the piece of equipment and rehire the men.

In another case, I was asked by a forest supervisor in Minnesota to conduct a 2-day workshop on the ecological value of old-growth forests. The supervisor had one district ranger who saw only dollar signs when he looked at the big, old trees and resisted anything that prevented him from cutting them as fast as possible. The problem was that the forest supervisor had established an old-growth committee throughout the entire forest to create a long-term, forest-wide management plan for the old-growth component of the ecosystem. For the committee to be effective, however, it had to include a representative from every district, but this particular ranger refused to assign anyone from his district to the committee.

Folks from throughout the forest and I spent the first day in a conference room, where I showed slides of how a forest functions above- and below-ground. I emphasized the role of large woody material, such as whole old-growth trees that fell onto the forest floor and ultimately decomposed into the forest soil. Again, this was a day of general discussion of people's perceptions and frames of reference. As such, it was riddled with abstractions for some of the people, including the ranger.

Knowing this, I took the group into the field the next day, where we explored the first day's abstractions in the concreteness of touchable examples. The ranger was quiet most of the time, appearing skeptical at best. Partway through the afternoon, however, he went up to the forest supervisor and said, "I never thought about a forest like this. I still don't understand everything I've heard, but I want to be on the old-growth committee myself."

This, again, was for me a totally unexpected outcome, but it illustrates the power of the mediation process to raise a person's level of understanding and thus their consciousness of cause and effect. When mediation is done in the way we practice it, the process protects the people's dignity and so makes it easier for them to change their thinking because they feel themselves to be in a venue of psychological safety, where their individual dignity is protected.

Inability to Transfer Experiences from One Situation to Another

Another major barrier to communication is the inability to transfer the outcomes of experience from one kind of situation to another. The potential ability to transfer results of experiences from here to there is influenced by the breadth of one's experiences. Every group represents a vast array of experiences, some broad, others narrow.

Experiential transfer, however, is critical to understanding how ecosystems and their interconnected, interactive components function, including the bridge between a community and its surrounding environment. It is also a necessary ability in resolving environmental conflicts, whereby potential outcomes can be projected to a variety of possible future conditions.

When participants cannot make such transfers for lack of the necessary frame of reference, they find many of the ideas to be abstractions, whereas others, with the required experience, feel them to be concrete examples based on their accumulated knowledge. This is where analogies are useful.

To make sure that my (Chris's) analogy will be understood, I ask the participants if they are familiar with the concrete example I propose to use in helping to extend their frame of reference to include the abstraction. If I am talking about the value of understanding how the various components of an ecosystem interact as a basis for the system's apparent stability, I may use the following examples:

1. What happens when just one part is removed? A helicopter crashed in Nepal some years ago, killing two people. A helicopter has a great variety of pieces with a wide range of sizes. The particular problem here was with the engine, which is held together by many nuts and bolts, each of which has a small, sideways hole through it so that a tiny "safety wire" can be inserted. The ends are then twisted together to prevent the tremendous vibration created by a running engine from loosening and working the nut off the bolt.

The helicopter crashed because a mechanic forgot to replace one tiny safety wire that kept the lateral control assembly together. A nut vibrated off its bolt, the helicopter lost its stability, and the pilot lost control. All this was caused by one missing piece that altered the entire functional dynamics of the aircraft. The engine had been "simplified" by one piece—a small length of wire.

Which piece of the helicopter was the most important at that moment? The point is that each part (structural diversity) has a corresponding relationship (functional diversity) with every other part. They provide stability only by working together within the limits of their evolved potential in biology or their designed purpose in mechanics.

2. What happens when a process is "simplified"? The newly elected mayor of a city, whose budget is overspent, guarantees to balance the budget; all that is necessary, in a simplistic sense, is to eliminate some services whose total budgets add up to the overexpenditure. This analogy represents the *symptomatic* approach of problem-solving, which is always simplistic. What would happen, for example, if all police and fire services were eliminated? Would it make a difference, if the price were the same and the budget could still be balanced, if garbage collection was eliminated instead?

The trouble with such a simplistic view is in looking only at the cost of, and not at the function performed by, the service. The diversity of the city is being simplified by removing one or two services without paying attention to the functions performed by those services. To remove a piece of the whole may be acceptable, provided we know which piece is being removed, what it does, and what effect the loss of its function will have on the stability of the system as a whole.

Once I am sure that the participants are following the analogy, then I can help them transfer the concept to the abstraction. As the principle of transfer becomes clear, the abstraction begins to take on the qualities of a concrete idea, and the barrier to communication is dissolved. Removing such barriers is important if the mediation process is to fulfill its greatest potential for safeguarding social-environmental sustainability for all generations.

Discussion Questions

1. Why is the need to be heard important in resolving a conflict? (See Biophysical Principles 1, 2, and 6 in Chapter 5.)
2. What are the basic elements of communication?
3. What are the barriers to communication? How do they work?
4. How could you make language real?
5. Why is it that people cannot transfer experiences from one situation to another?
6. Is there a particular question you would like to ask?

Endnotes

1. *The Wisdom of James Allen: As a Man Thinketh*. Laurel Creek Press, San Diego, CA, 2004, p. 22.
2. Pieter van der Zaag, Annette Bos, Andries Odendaal, and Hubert H. G. Savenije. Educating Water for Peace: The New Water Managers as First-Line Conflict Preventors. Paper prepared for the UNESCO-Green Cross "From Potential Conflict to Cooperation Potential: Water for Peace" sessions; 3rd World Water Forum, Shiga, Japan, March 20–21, 2003.
3. Ibid.
4. Van Mensvoort. Children's Dictionary Dumps "Nature" Words. February 4, 2009. http://www.nextnature.net/?p=3110 (accessed May 29, 2009).
5. Elaine Brooks. Eco Child's Play. http://ecochildsplay.com/2009/02/02/nature-words-dropped-from-childrens-dictionary/ (accessed May 29, 2009).
6. The foregoing three paragraphs on the words deleted and added to the Oxford Junior Dictionary are from Mensvoort, Children's Dictionary Dumps.
7. Richard Louv. *Last Child in the Woods: Saving Our Children from Nature-Deficit Disorder*. Algonquin Books, Chapel Hill, NC, 2008. 390 pp.

8. J. M. Healey, Early Television Exposure and Subsequent Attention Problems in Children. *Pediatrics*, 113(2004):917–918.

9. Quotes.net. Hellen Keler. http://www.quotes.net/quote/5990 (accessed June 16, 2017).

10. Don Costar. No Need to Change Grazing Practices [Letters to the editor]. *Reno-Gazette Journal*, Reno, Nevada. January 13, 1995.

8

Conflict Is a Learning Partnership

When parties can perceive and understand conflict as a learning partnership, it is much easier for them to ask their perceived adversary, "Will you please help me to sort out this problem?" And when they get the answer, it is rarely threatening. Through asking for help, they often find some terrified, repressed part of themselves looking back at them through their "adversary's" eyes.

In a sense, mediating the resolution of an environmental dispute is much like being a therapeutic counselor. To understand what this means, we recommend reading Chapter 14 in Corey's *Theory and Practice of Counseling and Psychotherapy*, from which the following quotation is taken: "Since counselors are asking people to take an honest look at themselves and to make choices concerning how they want to change, it is critical that counselors themselves be searchers who hold their own lives open to the same kind of scrutiny."[1]

It therefore falls on us, as mediators, to treat the mediation process as a learning partnership, which means that *we* must also be open to learning. If we are not willing to learn, we cannot teach because teacher and student are one and the same. Transformative mediation is thus an assembly of students and teachers who agree to learn with and from one another, with the mediator acting as an invited, impartial, guide throughout the process.

A Mediator Is at All Times a Guest and a Leader Simultaneously

A mediator must be a guest at all times in the process, as well as the leader. As such, Lynette and I, as mediators, must learn to lead. Before we can learn to lead, however, we must learn to follow. The act of leadership demands humility, whereas the outcome of leadership demands acceptance. This is one way of saying that we are responsible for our conscience, and everyone else is responsible for theirs. Thus, while we may, through our mediation process, help people to change their views, we cannot judge the way in which they do so.

Leadership Is the Art of Being a Servant

True leadership is concerned primarily with facilitating someone else's ability to reach their potential as a human being. Leadership comes from the heart and deals intimately with human values and human dignity. We must lead by example, as Francis Bacon noted when he said, "He that gives good advice, builds with one hand; he that gives good counsel and example, builds with both; but he that gives good admonition and bad example, builds with one hand and pulls down with the other."[2]

A leader knows and does what is right from moral conviction, usually expressed as enthusiasm, which causes people to want to follow with action. Essentially, a leader is one who values people and helps them transcend their fears so they might be able to act in a manner other than they were capable of doing on their own.

Leadership has to do with authority, which is control, or the right or power to command, enforce laws, exact obedience, determine, or judge. Two kinds of authority are embodied in this definition: that of a person and that of a position.

The authority of a person begins as an inner phenomenon. It comes from one's belief in one's higher consciousness, which acts as a guide in life when one listens to it: "As a man [woman] thinketh in his [her] heart, so is he [she]."[3] In contrast, a person who has only the authority of position may have a socially accepted seat of power over other people, but *power can exist only if people agree to submit their obedience to authority*. A person who holds a position of authority yet does not live from the authority within can only manage or rule as a dictator—through coercion and fear—but cannot lead from the heart.

A leader's power to inspire followership comes from a sense of authenticity because they have a vision that is other-centered rather than self-centered. Such a vision springs from strength, those universal principles that govern all life with justice and equity, as opposed to the relatively weak foundation of selfish desire. It is the authenticity that people respond to; in responding, they validate their leader's authority.

Management, on the other hand, is of the intellect and pays minute attention to detail, to the letter of the law, and to doing the thing "right" even if it is not the "right" thing to do. A manager relies on the external, intellectual promise of new techniques to solve problems and is concerned that all the procedural pieces are properly accounted for, hence the epithet "bean counter." Thus, a good manager is usually placed at a disadvantage when put in a position of leadership. Similarly, a leader placed in a managerial position is equally at a disadvantage because the two positions require vastly different skills.[4]

A good mediator, however, must be both an effective leader who guides the mediation process and an effective manager who keeps it running smoothly. By way of example, think of driving a herd composed of a hundred head of cattle.

There are three basic positions in driving cattle: point, flank, and drag. The person riding "point" is the leader, the one out front guiding the herd. The "flankers," or people riding along the sides of the herd, manage the herd by keeping it moving in the desired direction while preventing individuals from leaving the herd. "Riding drag" means to keep the cattle at the rear of the herd moving at a given pace while preventing individuals from dropping out of the herd. A mediator must metaphorically ride point, flank, and drag simultaneously because they are responsible for moving the whole herd safely from one intellectual place to another.

As a leader of the mediation process, one must be the servant of the parties involved. Servant leadership offers a unique mix of idealism and pragmatism.

The idealism comes from having chosen to serve one another and some higher purpose, appealing to a deeply held belief in the dignity of all people and the democratic principle that a leader's power flows from commitment to the well-being of the people. Leaders do not inflict pain, although they must often help their followers to bear it in uncomfortable circumstances, such as compromise. Such leadership is also practical, however, because it has been proven over and over that the only leader whom soldiers will reliably follow when risking their lives in battle is the one who they feel is both competent and committed to their safety.

Our first responsibility, therefore, is to help the participants examine their senses of reality, and our last responsibility is to say thank you. In between, we not only must provide and maintain momentum but also must be effective. Beware! Most people confuse effectiveness with efficiency. Effectiveness is doing the right thing, whereas efficiency is doing the thing as expeditiously as possible, although at times it may not be the right thing to do.

When the difference between effectiveness and efficiency is understood, it is clear that efficiency can be delegated, but effectiveness cannot. To us, effectiveness enables others to reach toward their personal potential through the mediation process. In so doing, we leave behind a legacy of assets invested in other people.

We are also responsible for developing, expressing, and defending the participants' civility and values. Paramount in any mediation process are good manners, respect for one another, and an appreciation of the way in which we serve one another. In this sense, civility has to do with identifying values, as opposed to following some predetermined process formula.

For a participant to lose sight of hope, opportunity, the right to feel needed, and the beauty and novelty of ideas is to die a little each day. For us to ignore the dignity of the mediation process, the elegance of simplicity and truth, and the essential responsibility of serving one another is also to die a little each day. In a day when so much energy seems to be spent on mindless conflict, to be a mediator is to enjoy the special privileges of complexity, ambiguity, diversity, and challenge.

As auto manufacturer Henry Ford once said, "Coming together is a beginning; keeping together is progress; working together is success."[5] In the end, it is the collective heart of the people that counts; without people, there is no need for either leaders who mediate or mediators who lead. Lao Tzu, the Chinese philosopher, said, "A leader [mediator] is best when people barely know he exists, when his work is done, his aim fulfilled, they will say: we did it ourselves."[6] Such is servant leadership, and such is our goal. To achieve our goal, we must make the mediation process safe enough for people to trust it and willingly give up their hidden agendas.

Hidden Agendas

It is a mediator's responsibility to make the process safe enough that all hidden agendas are placed on the table; otherwise, they can destroy the essence of mediation. A hidden agenda occurs when a person holds back the information about what they really hope to gain from the process. Hidden agendas vary, and only two are discussed here as examples. The first has to do with the hidden agenda of a mediator and the second with that of a participant.

Mediator

Some years ago, I (Chris) was part of a national committee dealing with grazing fees on public lands. Our charge was to advise the secretaries of both the Department of the Interior and the Department of Agriculture on how to structure the grazing fee incentive program.

Fairly early in the two-and-a-half-day meeting, the interior secretary joined us for most of a day. It soon became clear that he wanted a particular outcome, and that he had a definite timetable. This was bad enough in and of itself, but what was worse was that the two mediators knew what he wanted and were doing their best to covertly push us in that direction. I had the distinct feeling, as did others, that the mediators were trying first and foremost to please the interior secretary by getting as close as possible to what he wanted as quickly as possible.

The mediators' hidden agenda became clear on the first day. Whenever someone expressed their feelings about the issue of grazing and range condition in general, a necessary part of the process, the mediators did their best to cut that person off and return to the strictly structured, carefully controlled, predetermined mediation format. We were not allowed to deal with our senses of value, either individually or collectively. There were 30 of us who, for the most part, were complete strangers when the meeting was first called to order; thus, we never really got to know one another.

I, as a participant, ended up not trusting the mediators' intentions and not liking the way I was treated. Consequently, I would not choose to work with them again. They appeared to be strictly product oriented in a self-serving way. Although I do not know what they hoped to gain, my intuition at the time told me it was further employment.

Participant

I (Chris) was once asked to participate, as an independent observer, in a consensus group. At least 30 points of view were represented because at least 30 people, in addition to the mediator and myself, were present. During the 2-day meeting, I interpreted three general "collective" views, two of which represented a long-standing battle over whether to cut a particular city's water catchment. Because I knew nothing about the conflict, even though it had been alive for some years, I had no vested interest in it and could therefore see the collective views. Let us examine my interpretations of them, one at a time.

View 1: A most sincere, elderly lady, who had lived in this city all her life, had been told in the third grade that the city's water catchment, covered with virgin old-growth forest, was her national heritage and would never be cut. Now, she found people from a land management agency clear-cutting "her water catchment," and she felt betrayed. Where the third-grade teacher got the notion of an inviolate national heritage is moot. The lady, joined by her son, thought the land management agency should cease and desist all cutting and road building in the water catchment forever. On this she was adamant.

View 2: The conservation groups that were represented were unanimously opposed to further logging and roading in the water catchment because the virgin old-growth forest created and protected the pure quality of the city's water supply.

View 3: The people from the land management agency saw the old-growth timber as an economic commodity that had to be cut and milled or there would be an irreparable economic loss because the old-growth forest would fall down and rot, an unthinkable economic waste.

All three views, each with a stake in the water catchment, played the consensus game with a hidden agenda. The hidden agenda each side was trying to conceal from the others, while acting innocently open minded, became obvious as the mediation process unfolded. Although the hidden agendas were never admitted, much less openly laid on the table, they were covertly defended whenever someone got too close to the truth—a sure sign that all was not as it appeared.

View 1: The elderly lady and her son had become rather prominent as distributors of a small newsletter to the group of conservationists interested in saving the water catchment's old-growth forest. If the lady and her son won their point of view, they would disappear into the oblivion from which they came; with the issue resolved, the other folks would turn to new issues. Thus, whenever reconciliation seemed possible, the son categorically refused to accept anything that had the appearance of moving the problem toward solution. His hidden agenda seemed to be to keep the issue alive and thereby forestall the feeling of rejection through loss of importance and thus loss of identity.

View 2: The conservationists were committed to saving the old-growth forest (trees). Each time the people from the land management agency conceded a point that would benefit water quality but not save the trees, the conservationists had to find a new point from which to argue, one that sounded valid with respect to clean water and did not mention trees.

View 3: The people from the land management agency were committed to cutting the timber for economic reasons. They thus submitted to the procedure, but with the knowledge that the authoritative position and final decision was on their side.

Where did we go from here? First, each person was right from their point of view, from their interpretation of the data. Second, no one in the room really understood consensus. Consensus does not mean something will be enacted; it means that the parties agree to agree on something. And, the agreement the participants ended up with was that something needed to be done, which is where they started.

The mission was doomed to failure because *no one* disclosed their real agenda. Why not? They withheld their agendas because there was a lack of trust in the process. After all, the mediator was an employee of the land management agency represented in the dispute, in addition to which the agency both paid for and hosted the session. Without being an impartial third party who had earned the participants' trust, the mediator could neither lead nor teach, even by example.

Rethinking the Use of "Consensus"

Some people consider the term *consensus* to be negative. It means that *everyone agrees* with the direction a group is taking, which is often not the case. Dr. Hans Bleiker, founder of the Institute for Participatory Management and Planning, suggested that the term *informed consent* must be considered the ultimate outcome of our mediation efforts, rather than seeking "consensus" from a group of participants. He defined informed consent as "the grudging willingness of opponents to go along with a course of action that they are actually still opposed to."[7] It means that, as mediators, we hope the participants will accept the decision that is made, even if they do not agree with it 100 percent, but nevertheless feel they can live with the decision and support it.

There are going to be times, however, when consensus is just not possible. To proceed in those cases, one could obtain the consent of each participant to move forward, despite their inability to reach agreement, because it is best to move forward rather than make no decision or take no action. The "Bleiker Life Preserver" is a four-step thought process used to guide a group toward informed consent. In brief, it consists of four statements:

1. This is a serious problem (or an important opportunity) that must be addressed.
2. This group (or agency) is the right entity to address it. In fact, it would be irresponsible for you/them, with the mission that you/they have, not to address it.

3. This approach is reasonable, sensible, and responsible.
4. We do listen; we do care. (Everyone has heard your concerns, and we do care about making the right decision.)[8]

Mediator as Teacher

Whether one realizes it or not, a mediator is a teacher in that learning is a change in behavior as a result of an experience, which is the purpose of transformative mediation. Most people who seek mediation have fairly definite ideas about what they want out of the process, and they will learn from any activity that tends to further their purposes. Because previous experience conditions a person to respond to some things and to ignore others, it is imperative that the mediation experience is made relevant to the desired outcome, namely, that the parties learn to reconcile their differences and work cooperatively with one another. To accomplish this, some elements of learning must be understood.

The Foundation of Learning

If an experience is to challenge the learner, it requires involvement with feelings, thoughts, and memory of past experience, as well as necessitating physical activity. In addition, a physical activity is more effective than one in which all a person has to do is deal with abstractions or commit something to memory. Each person approaches a task with preconceived ideas and feelings and may have these ideas changed for a variety of reasons as a result of experience. The learning process may therefore simultaneously include the following: verbal, conceptual, perceptual, emotional, and problem-solving elements.

Learning is multifaceted in another way. While learning a subject at hand, people learn other things as well. They develop positive or negative attitudes about mediation, depending on their experience with the process. They may, for example, learn greater self-reliance under the guidance of a skillful mediator. Such incidental learning may have a great effect on a person's total development, hence the transformative nature of mediation.

The effectiveness of learning is based on a person's emotional response in that learning is strengthened when accompanied by a pleasant feeling and diminished when associated with an unpleasant one. It is thus better to tell the parties involved in the mediation process that their problem, although difficult, is within their capability to understand and resolve. Having said this, however, it is incumbent on us, as mediators, to keep our word. Therefore, whatever learning situation is encountered in the mediation process, it must contain elements that affect the parties positively and produce such feelings as self-worth, success, freedom, clarification, and empowerment, all of which enhance learning.

Part of the foundation of learning is primacy, the state of being first, which often creates a strong, almost unshakable, impression. This means that what we teach through the mediation process must be right the first time. For the parties, it means that the learning must be right the first time because "unteaching" and "unlearning" are far more difficult than teaching and learning. It is thus critical that the first experience be positive and functional for everything that follows is predicated on this first experience. Part and parcel of this first experience is the notion of intensity because a vivid, dramatic, or exciting experience teaches more than a routine or boring one.

How People Learn

Learning comes initially from perceptions directed to the brain by one or more of the five senses: sight, hearing, touch, smell, and taste. Psychologists have determined experimentally that "normal" individuals acquire about 75 percent of their knowledge through sight, 13 percent through hearing, 6 percent through touch, 3 percent through smell, and 3 percent through taste.[9] Learning occurs most rapidly, however, when information is received through more than one sense.

Nevertheless, real meaning can only come from within a person, even though the sensations evoking these meanings result from external stimuli. People therefore base their actions on the way they *believe* things are.

Factors Affecting Perception

Learning is a psychological problem, not a logical one. Therefore, the mediator must organize the process to fit the psychology of the participants. As long as a person feels capable of coping with a situation, each new experience can be accepted as a challenge; if, on the other hand, a situation seems overwhelming, the person usually feels unable to deal with it and perceives a threat. Mediation is thus consistently effective only when those factors influencing perceptions are recognized and taken into account, as positively as possible.

Among the factors affecting perception are (1) basic necessity; (2) self-concept; (3) timing, opportunity, and time; and (4) recognizing an element of threat. *Basic necessity* is a person's need to maintain and enhance the organized self. The self is complete in that it is a physical and psychological combination of a person's past and present experiences, as well as future hopes and fears. A person's most fundamental, pressing necessity is perceived to be perpetuating this identity of *self*, which in turn affects all perceptions.

Just as the food one eats and the air one breathes become part of the physical self, so the sights one sees and the sounds one hears become part of the psychological self. We are psychologically what we perceive.

As a person has physical barriers that prevent dangerous things from harming the physical being, such as flinching from a hot stove, so a person has perceptual barriers that block those sights, sounds, and feelings thought to pose a psychological threat. Helping people to learn thus requires finding ways to aid them in developing different perceptions in spite of their dysfunctional coping mechanisms. Because a person's basic necessity is felt to be self-maintenance, the mediator must recognize that anything asked of a party that may be interpreted as imperiling the self will be resisted and denied. Here, a thought by Henry Wadsworth Longfellow is apropos, "If we could read the secret history of our enemies, we should find in each man's life, sorrow and suffering enough to disarm all hostilities."[10]

Self-concept, or how one pictures oneself, is a powerful determinant in learning. A person's self-image, described in such terms as *confident* or *insecure,* has much influence on one's total perceptual process. If a person's experiences in the mediation process tend to support a favorable self-image, one is more likely to remain open to subsequent experiences. If, on the other hand, a person has negative experiences that threaten their self-concept, there is a tendency to reject additional participation.

The people in the mediation process who view themselves positively are less defensive and more readily internalize and assimilate their experiences. But, those with negative self-concepts activate their psychological barriers, which tend to keep them from perceiving and may actually inhibit their ability to implement that which is perceived in a functional manner.

Timing, opportunity, and time are necessary to perceive and learn. Learning depends on perspective and its perceptions, which precede those perceptions to be learned. Timing is thus important because a person may not be ready to learn certain things without prior experience. Assuming the timing is right, one requires both the opportunity and the necessary time to accommodate the experience of learning. In addition, the amount of time necessary to learn a given thing differs from person to person.

Finally, *fear* adversely affects one's perception by narrowing the perceptual field. People confronted with a threat tend to focus their attention on the perceived danger, which reduces their field of vision to a fraction of its potential. Anything a mediator does that is interpreted as threatening makes the already frightened person less able to accept a new experience by adversely affecting their physical, emotional, and mental faculties.

Insights

Insights involve grouping perceptions into meaningful wholes or systems thinking. *Evoking insights is a mediator's main task*; therefore, it is essential to keep each person constantly receptive to new experiences and to help each person realize and understand how a given piece relates to all others in the formation of patterns. Understanding the way in which each piece may affect the others and knowing the way in which a change in any one may affect changes in all others is imperative to true learning. Although insights almost always occur eventually, effective mediation can speed the process by teaching the relationship of perceptions as they occur, thus promoting the development of transformative insights in the mediation process.

Motivation

People in a mediation process, like all other workers, want a tangible return for their efforts. If motivation is to be effective, the participants must believe their efforts will be positively rewarded. Whether such rewards are the furtherance of self-interest or group recognition must be constantly apparent during the mediation process.

Although many lessons with obscure objectives will pay handsomely later, a person may not appreciate this immediately. If motivation is to be maintained, therefore, it is important to make the participants aware of applications that are not immediately apparent.

Mediators often inadequately appreciate the desire for personal security and comfort. All participants want secure, pleasant conditions and states of being, even under the most trying of circumstances. If they recognize that what they are learning can promote this goal, their interest is easier to attract and hold.

Along these lines, psychologist Abraham Maslow's hierarchy of needs is helpful in understanding this point. Although Maslow's hierarchy of needs is given in the shape of a pyramid, we can visualize the same concept by picturing a stepladder with a very wide base and a narrow top. Beginning with the first rung from the bottom are the basic physiological survival needs, such as air, water, food, and shelter. On the second rung are the needs for safety, security, stability, structure, order, limits, and law. The third rung deals with belongingness and love, which express themselves in the need for roots, origin, being part of a social group, and having a family and friends. The fourth rung encompasses self-esteem, which manifests itself in the need for strength, mastery, competence, self-confidence, prestige, status, and dignity. The fifth and highest rung is self-actualization, which is the inner, driving need to become all that one can be by knowing truth, justice, beauty, simplicity, and perfection.[11]

Notice how far up the ladder (the fourth rung) one must go before dignity is mentioned. Notice also how many external, product-oriented needs must be met before a person feels in enough control of life's conditions to "afford" a sense of dignity.

Fortunately, there is within each person engaged in a task the belief, however small, that success is possible under the right conditions. This belief can be a most powerful motivating force. A mediator can best foster such motivation by introducing perceptions based solidly on experiences that are easily recognized as achievements in learning.

A majority of people seem to have trouble transferring abstract concepts from one situation to another, as previously discussed. Therefore, concrete examples are important and can be supplied through the use of appropriate analogies, which help participants transfer a commonly understood experience to a foreign situation. For example, a human community can be used to explain how an ecosystem functions. Once the basic principles are under-stood in a human context, they can be transferred to a forest, grassland, or an ocean, always coming back to the human community as a touchstone. In addition to using good analogies, a wise mediator constantly relates the lesson's objectives to the participants' intentions and necessities and thereby builds on the participants' natural enthusiasm.

The relationship between a mediator and participants has a profound effect on how much the participants learn and change. Lynette and I (Chris) must create as safe and gentle an environment as possible through our own demeanor in order to enable participants to help themselves.

The following generalizations about motivated human behavior may be helpful:

1. Work is as natural to people as are play and rest. Work that is a source of satisfaction will be voluntarily performed, but that which is per-ceived as a form of drudgery or punishment will be avoided if possible.

2. A person will exercise self-direction and self-control in the pursuit of goals to which they are committed.

3. A person's commitment to their goals is directly related to the per-ceived reward associated with their achievement.

4. A reasonably functional person learns, under the right conditions, to both accept and seek responsibility. Ambivalence and shirking responsibility are not inherent in human nature, but rather are usu-ally consequences of dysfunctional experiences during childhood and negative, unsafe experiences in life.

5. The capacity to exercise a relatively high degree of imagination, ingenuity, and creativity in the resolution of common problems is a widely distributed human trait.

6. Under the conditions of modern life, the intellectual potentialities of the "average" person are only partially used.[12]

Lynette and I accept these assumptions and see vast, untapped potentiali-ties in participants. The raw material lies waiting; its release is partly in our hands and partly in those of the participants.

The Fallacy of Rescuing

Over the years, I (Chris) have been in many situations where either an indi-vidual person, such as the previously discussed fatalist, or a whole group wanted to be rescued. When someone wants to be rescued, they inevitably want a quick fix, with me, the mediator, doing all of the work.

I have experienced two major problems with this notion. First, the person who ostensibly wants to be rescued will continually fight any effort to be rescued. To this person, wanting help to change sounds good, but they are not ready to give up the long investment in their current situation. They are getting some kind of value out of it, even if unconsciously.

The second problem is that, even if I could and would rescue someone, my efforts would be of no value. For example, while in Japan in 1992, I spent much time looking at forestry problems and discussing what I saw with Japanese foresters, prefecture mayors, and others. It was clear that the plantations of larch trees, put in by the Americans following World War II, were not only a mistake but also sick and needed to be replaced with native forest.

After I got home, a Japanese gentleman wrote to me and asked me to plan a forest for a particular area that could be grown for a thousand years. I told him, "No." I said that if I, as an American, were to plan a forest for the Japanese, I would be giving the Japanese an American forest planned in America by an American; it would be of no cultural value to them because it would be my forest, planned for me, and based on my own sensibilities. I did say, however, that I would *help* the Japanese plan a forest in Japan for themselves. Then, it would be Japanese planning their own forest, with their own species, for their own culture. Only then would the forest be of any value to them.

The point is that each person or group of people must struggle with and through their own processes if they are to derive anything of value from them. Even if I could rescue someone (go through their growth process for them), I would not. To do so would be stealing their struggle to grow and whatever value they would have derived from it. Besides, that which is not earned is casually tossed aside because the person, not having earned it, finds in it little or no value. It is thus a mediator's ever-present responsibility to define—through clear boundaries—their relationship with the participants while mediating the resolution of a dispute.

A Mediator's Role in Participant Relationships

Helping achieve good human relations is one of our (Chris's and Lynette's) basic responsibilities. To achieve such relations, we must consider the following points:

1. People gain more from wanting to resolve their own issues and learning in the process than from being forced to participate through such means as court directives.

2. Participants tend to feel secure when they know what to expect or what is going to happen. When they understand the benefits of what is taking place, they are more willing to move forward.

3. Each individual within a group of participants has a unique personality, which we must constantly consider. If, however, we limit our thinking to the group as a whole, without considering the participants as individuals, our efforts are directed toward an average personality, which fits no one.

4. If we give sincere praise or credit when due, we provide an incentive to strive harder. By the same token, insincere praise given too freely is valueless.

5. If a participant is gently briefed, in private, on erroneous assumptions and told how they might correct them, progress and accomplishment follow.

6. It is vital that our philosophy and actions are consistent. If a situation, such as being allowed to interrupt and speak out of turn, is acceptable one day but not another, participants become confused.

7. No one, including participants, expect us to be perfect. Nevertheless, the best way for us to win the trust and respect of participants is to honestly admit mistakes. If we try, even once, to bluff or cover up, the participants sense it quickly, and our behavior destroys their confidence in both the mediation process and us. Therefore, if in doubt about some point, we must freely admit it.

These are but a few of the many attitudes and reactions that help establish the kind of mediator-participant relations that promote resolving conflicts through effective learning. To accomplish this kind of learning, however, one must be present in the moment, every moment.

Mediation Means Total Participation

Being present—here and now—is the only way to participate in life, and effective mediation *requires* total participation. But, neither Lynette nor I (Chris) can be present if we are thinking about either the past or the future. It is therefore critical that we are totally focused on the present and also keep the parties focused totally on the present during the entire process.

What is so important about being present? Have you ever noticed that the present is seldom quite right or seldom seems good enough? Yesterday is past and gone; you cannot change it—opportunities foregone are opportunities lost. Tomorrow is not here, and you have no idea, despite your aspirations, hopes, expectations, and predictions, what it will bring. In reality, tomorrow is something that is always coming, yet never arrives. The present, the here and now, is all anyone ever has. This is *everyone's eternal reality*.

Being present, in the sense of being mentally here, now, is a difficult concept because there is no word that means mental presence, as opposed to simple physical presence. Let us examine what being present means.

Suppose that, while driving, you are thinking about (remembering) the past, say last year's vacation. It was your first trip to the Bahamas, and you had a very rough flight. The unexpected storm really frightened you, but once in the Bahamas, you had a marvelous time. You are flying to the Bahamas next month for your long-anticipated vacation, and you begin thinking about how much fun you had last year and you expect this year. Suddenly, out of nowhere, you have a vivid flash of last year's flight and become afraid that next month's flight might be the same. In all of your reverie, you are either in the past or in the future, and now you are jerked into the present as your car sputters and you coast to the side of the road out of gas. You were so busy thinking about your upcoming flight and vacation that you missed the gas station, despite the fact that you knew the gas tank was almost empty.

Being fully conscious in the present is important because fear is a projection of a past experience into the future. You cannot, therefore, be afraid in the present, in the here and now, because the present moment cannot be projected into the future; only the past can be projected into the future. What you really fear is the future, the fear of the unknown, the fear of being afraid—the fear of fear. Struggling to keep yourself in the present is thus a conscious choice you must make each time fear raises its ugly head.

Another reason for being in the present is to allow someone else in the present with you. I (Chris) sometimes think, for example, that I know what someone is going to say or how they are going to say it, whatever it is. I find this especially true if I do not want to hear either the answer or the tone of voice I *know* I will get. If I do not keep myself in the present, if I expect the person's old pattern of response, then *I limit* their ability to respond differently in the here and now because I simply do not hear it. I hear what I expect to hear, which justifies my expectation and imprisons both the other person and our relationship somewhere in the past. It also imprisons me in the past because all I accept, and therefore all I hear, is old business.

If we (in the generic) are not present with one another, our attempts to communicate can become frustrating experiences of talking either at or past one another, because the present, the here and now, is all we have. It is thus a mediator's responsibility to bring the parties back to the present whenever they stray unnecessarily into either the past or the future. Such presence is a prerequisite for detachment and equanimity throughout the mediation process.

Detachment and Equanimity

Detachment from an outcome is total acceptance of what is without any desire to have something else, which is a critical concept in transformative mediation. Detachment is checking our ego at the door as we come into the room. This is, at best, difficult to learn, and I (Chris) have consciously struggled with it over the decades.

When I (Chris) was younger, I was deeply upset by the clear-cut logging of the ancient forests in the Pacific Northwest, where I grew up. I would argue long and loudly about the need to save them and the greed and stupidity of those who wished to liquidate them. I tried to convince anyone and everyone that the forests needed to be saved. I was so rabid about my point of view being the right one that few people cared to listen unless they already agreed with me. Consequently, I became frustrated, cynical, and self-righteous, all of which only made matters worse. I became enraged at the *greedy bastards who were clear-cutting my forest*, but I never thought to ask them how they felt about the forests they were liquidating.

One day, as I was giving a passionate speech on the need to "preserve" the ancient forests of the Pacific Northwest, I suddenly felt the sword taken from my hand and a sense of peace come over me, a sense that was immediately reflected in the audience. Several people came up to me later and said they had never thought about it that way, and that what I said made sense. It was then I realized that to speak for the forests or for anything else, I had to change—not the people in the audience, but myself. If I wanted people to listen, it was incumbent on me to change, to say what I had to say in a way that would allow them to hear. But how? I did not know how.

A few weeks later, I saw the movie *Gandhi*. Then, I read a couple of biographies about Gandhi in which he was often quoted, and through his writings, he gave me the answer. I had to detach myself from the outcome, a truly difficult task.

If Gandhi was correct, in detachment lay acceptance of the outcome. Expectation is the attachment, the vested interest in the outcome, because people with expectations see themselves as the ones possessing the means of achieving the right and justifiable result. If, on the other hand, one acts willingly out of duty to a Higher Authority, one can act with detachment because the Higher Authority is acknowledged as the only one with the wisdom to justly govern the outcome.

If I am detached, I have no vested interest in the results of a given process, and I can treat all sides, all points of view, and all possible outcomes with equanimity. Equanimity is the kernel of peace in detachment just as surely as anxiety is the kernel of agitation in attachment.

For example, a person who has worked passionately for a cause may suddenly have the insight that passion placed before principle is a house divided against itself, which so divided cannot long stand. Because of this new understanding, they now become focused on the principle, as a process, and become detached from the passion—the desired result. The reaction of their peers most often is, "How can you give up the cause? We've believed in it for so long."

Attachment to the cause has for these people become life itself, their very identity, as discussed in the Hidden Agendas section. Therefore, even as they ostensibly fight to "win," they cannot afford to win because if they were to actually resolve the issue at the heart of their cause, they would have to find a new identity, something most people are loath to do.

If a mediator is truly detached from the outcome, they will find equanimity to be their touchstone. Equanimity, the outworking of detachment, is reflected in the calm, even-tempered, and serene personality of one who is simply open to accepting what is. Such a person can perform mediation without the need or the expectations of approval or a predetermined outcome. Such a person acts out of peace.

In turn, the peaceful action allows others to see an alternative way of perceiving something because no one is trying to convince them of anything. They are given the ideas and the space to consider them. Then, if they so choose, they can change their minds in privacy while retaining their dignity intact.

The one who is detached is part of the principle and is therefore part of the resolution or the transcendence of the problem. On the other hand, one becomes part of the problem when one is attached to a point of view and its necessary outcome. Detachment and equanimity serve to make the mediation process safe enough to permit the expression of anger, which can be defused because I (Chris) keep it focused on me, where it has no effect. Why? This is because I do not personalize it.

As a Mediator, You Must Be a Sieve, Not a Sponge

Anger, as previously stated, is fear violently projected outward from a person onto another person or an object. But, I (in the generic) am *not* angry for the reason or with the person or thing on which I am focused. I am angry at myself for being afraid of circumstances and therefore feeling a loss of control, which has nothing to do with the person or thing at which I level my anger.

That notwithstanding, people must be able to vent their anger in safety during the mediation process because the anger is there, and it will go somewhere. It is therefore important that a mediator keep the anger focused on themselves, but as though they are a sieve through which the anger simply passes because it is nothing personal.

For example, when a participant aims their anger at another participant, I (Chris) redirect that person's anger at myself by saying, "Excuse me, but I believe you were talking to me." The person can thus dissipate their energy, feel they have been heard, have their fears validated, and become more receptive to other data and ideas. At the same time, I keep the mediation environment as safe as possible by preventing another participant from being directly attacked in front of everyone else. Because I know the individual is not angry with me, their anger has no effect on me. By analogy, it is like a gift I do not want, so I refuse to accept it.

To illustrate, I remember meeting with the radical environmental group Earth First! in Arcata, California, some years ago. They had asked me to speak about old-growth forests and sustainable forestry. During my presentation, I pointed out that one must be careful not to become what one is against in confronting a perceived injustice with violent civil disobedience.

Some of the people did not understand what I meant, so I gave the example of combating violence with violence. "If I am treated with violence," I said, "and I therefore *react* with violence, how am I any different than my opponent? I am not. I have in attitude, behavior, and tactics become what I am against; I have become as my opponent."

At this point, the most militant individuals arose, yelling and swearing at me with such vehemence that my wife, who was in the audience, became terrified for my safety. Their rage was electrifying. Had I stopped any of it, taken any of it personally, or in any way defended myself against it, the situation would easily have gotten out of hand and become dangerous. Instead, I let it pass through untouched, and the militants all stormed harmlessly out of the room and did not return.

Because I did not *respond* to the anger, a number of people came up to me after I was through speaking. To a person, they said that they had not thought of violence in the way I had presented it and would reconsider their stance. If only one person had said that, it would have been well worth the blast of rage leveled in my direction.

If a mediator ever makes the mistake of taking someone's anger personally, of internalizing it, they become like a sponge soaking it all up. Being a sponge for another's anger not only is detrimental to the mediator emotionally but also causes them to become a major problem in the mediation process. At this point, they can no longer function as a mediator because they now have a vested interest in (can no longer be detached from) the outcome.

As the mediation leader, I (in the generic) must bear, unflinchingly, all the abuses that the parties normally hurl at one another. In effect, a person, such as a mediator, who serves the people must pass the tests described in the

eulogy that Senator William Pitt Fessenden of Maine delivered on the death
of Senator Foot of Vermont in 1866:

> When, Mr. President, a man becomes a member of this body he cannot
> even dream of the ordeal to which he cannot fail to be exposed;
>
> of how much courage he must possess to resist the temptations
> which daily beset him;
>
> of that sensitive shrinking from undeserved censure
> which he must learn to control;
>
> of the ever-recurring contest between a natural desire for public appro-
> bation and a sense of public duty;
>
> of the load of injustice, he must be content to bear, even from those
> who should be his friends;
>
> the imputations of his motives; the sneers and sarcasms of ignorance
> and malice;
>
> all the manifold injuries which partisan or private malignity, disap-
> pointed of its objects, may shower upon his unprotected head.
>
> All this, Mr. President, if he would retain his integrity, he must learn to
> bear unmoved, and walk steadily onward in the path of duty, sustained
> only by the reflection that time may do him justice, or if not, that after
> all his individual hopes and aspirations, and even his name among men,
> should be of little account to him when weighed in the balance against
> the welfare of a people of whose destiny he is a constituted guardian
> and defender.[13]

Such is the price of leadership: to be the keeper of everyone else's dignity.

As a Mediator, You Are the Keeper of Each Participant's Dignity

As I (Chris) understand dignity, its emotional foundation rests on the perceived ability to make choices, which in turn provides a sense of hope. As
mediator, I am the keeper of each participant's dignity, which means I will
protect their dignity so they can protect mine. Protecting one another's dignity is tantamount to making and keeping mediation as safe and gentle as
possible. Being the keeper of the participants' dignity means there is no
blame or guilt, only an opportunity to think differently.

I spoke some years ago at an annual banquet for the Florida Audubon
Society in Tallahassee. The day after I spoke, I was on a panel with the supervisor of one of the national forests in the southeastern United States. His

opening story exemplifies dignity as the ability to make choices and having some things of value from which to choose:

> "I am concerned about what we are doing to our forests," he began. "I am concerned about what we are saying to one another. I think about it all the time. My nine-year-old son came home from school the other day in tears. 'What's wrong?' I asked."
> "Dad," he replied, "my teacher said you're destroying the forest. You're cutting it all down. What's going to be left for me when I grow up?"
> The forest supervisor was trembling with emotion. I could feel his terrible pain, and because of his pain, I am not sure he was even aware of the audience as he spoke.
> "I'm concerned," he said, "about what we're teaching our children in school. I'm concerned about what we're telling them about their future. I think about it all the time."

The forest supervisor's young son was asking his father, Where is my choice? Where is my hope? What will be left for me? What will give me a sense of well-being, of dignity, when I grow up?

And yet, we, in the United States, are teaching the children of countries we smugly deem to be "underdeveloped," "developing," or "third world" that they are somehow "lesser" or "inferior" human beings than our children and we are. I say this because, of the several facets reflected in the term *development*, we have chosen to focus on a very narrow one: the growth of personal materialism—a *standard of living*—through centralized industrialization, which we glibly equate with social "progress" and "economic health."

I have over the years worked in a number of countries without giving much thought to the notion of "developed" versus "developing" or "underdeveloped," as some would put it, although I have spent time in each such area. During a trip to Malaysia some years ago, I was profoundly struck by the arrogance and the narrowness of such thinking. What we are really talking about is a degree of *industrialization*, and that is a different issue.

That said, Malaysia is the only place where I have ever heard the people refer to their own country as "developing," as though they were lesser than developed countries and must somehow "catch up" to be equal—a notion of inferiority their children are learning indirectly. Moreover, they often apologized to me for being *lesser human beings*, prior to asking my permission to ask me a question, and that included apology by a university professor.

Our subjective feelings and human values govern our choices. Our so-called rational, objective intellect or our scientific knowledge or technological advances do not govern them. Remove the sense of choice that has to do with one's determination of one's own destiny and you remove hope as well, and human dignity withers.

Because dignity is closely tied to being in control, humiliation is dignity's greatest enemy. Let us return to the government meeting I (Chris) spoke about in the discussion of hidden agendas. As I said, the facilitators were intent on meeting the interior secretary's agenda for grazing on public lands. To meet that agenda, they had to control the length of time each person spoke, especially on the last day. To do this, they made a big red paper flag with large yellow letters on it: BS (bullshit). Then, when they thought someone had spoken long enough and wanted to hurry the meeting along, one of them would wave the BS flag from the front of the room in an attempt to humiliate the speaker into silence. What kind of example did they set? Was it a safe place for people with different values to come together? What did they do to the participants' trust and dignity?

Mahatma Gandhi once said, "It has always been a mystery to me how men can feel themselves honoured by the humiliation of their fellow beings."[14] I (Chris) have been humiliated in my life. I never felt that my loss contributed to anyone else's honor. I always felt robbed by humiliation because the person who humiliated me was trying to extract my obedience, make me lesser—always lesser—to gain my submissiveness to their control. And I do not want to be controlled; I want to be respected. I, like everyone else, need love and help, not criticism, judgment, and humiliation.

A young man taught me a wonderful lesson about humility and human dignity, a lesson I will always remember. It was a cold winter, and most of us kept fires going in our woodburning stoves when at home. One day, this young man's wife cleaned the ashes out of their stove. She put them into a paper bag and put the bag in the back of their pickup truck. What she did not know, however, was that some of the ashes were still hot. In fact, they were hot enough to catch the bag and other things in the back of the truck on fire and severely damage the interior of the pickup's canopy.

"What did you say to your wife about burning up your truck?" I asked, as if it were any of my business in the first place.

"Nothing," he said. "She was already humiliated. Why should I add to it? After all, she didn't do it on purpose."

That was not the response I had anticipated. People often get a lot of mileage out of other people's mistakes and proceed to flog them with embarrassment.

What would the world be like if we were all as kind and thoughtful as that young man? What would the world be like if we were all the caretakers of one another's dignity? Do you think we can resolve the problems for which the lack of human dignity is an obstacle in the healing of our global society? As a mediator, you have a chance to begin doing just that, healing society, but you must have a beginner's mind to see the oft-hidden opportunities.

Have a Beginner's Mind

To approach life with a beginner's mind, a mind simply open to the wonders and mysteries of the universe, is a gift of Zen. A beginner, unfettered by rules of having to be something special, sees only what the answers might be and knows not what they should be. The one who deems themselves to be an expert, on the other hand, is bounded by the rules that govern being an expert. Such a person acts as though they are special, the one who knows what the "correct" answer *should* be, yet is too often blind to what other answers might be. The beginner is free to explore and to discover, while the self-appointed expert grows rigid in a self-created prison.

Two women and one man on three different occasions, years apart, are excellent examples of the beginner's mind. I (Chris) often use a simple, fun exercise that requires nothing more than six wooden matches or six tooth-picks to help people understand that their imagination is as bounded as their blind acceptance of social convention and as free as their willingness to reach beyond such convention in seeking their soul's creative eye.

The instructions are simple: Sit at the table and make four equilateral tri-angles out of the six matches or toothpicks *without* crossing one over another. Rarely does a person succeed because to accomplish this feat on the sin-gle dimension of a table's flat surface seems impossible, which their mind quickly tells them, even as they struggle not to accept it. They think it must be possible because I told them to do it, but they cannot figure out how and eventually give up. There are, however, at least two ways to solve this prob-lem, one of which had been unknown to me.

The way I had learned to solve the problem was to make one triangle on the table's flat surface and then stand the other three bits of wood upright within the one, thus encompassing more than a single dimension. The other way is to break a match or toothpick and arrange them appropriately on the table's surface.

This first time I saw the problem solved this way was at a workshop I was conducting to help wildlife biologists look beyond professional convention for answers to their management problems. During the workshop, one of the biologists came to me and said that his wife, who had accompanied him to the meeting, was interested in what I was talking about and asked if she could join us. "But, of course," I said.

With the lady sitting at the table, I gave my usual instructions and then simply watched what happened. While all the men arranged and rearranged their matches to no avail, she put hers on the table and sat looking at them. Suddenly, a tiny smile crept over her face. Picking up the matches, she laid one down at an angle. Then she deftly broke one in two, laying each half across from the other on each side of the middle of the first match. Finally, she arranged the remaining four in a square to close the exposed sides. Although not perfect, she had four triangles!

"Well I'll be damned," was all I could say as she beamed at me from across the table.

Over the years, I continued giving my original instructions, wondering if anyone else would break a match. Finally, after more than two decades, a sixth-grade teacher looked at her matches for about 30 seconds, then looked up at me and asked, "Can I break the matches?"

"Yes," I answered, and then I asked her, "How did you figure that out so fast?"

"Well," she replied, "any time I'm given limits, the first thing I do is check them to see if there's an alternative."

What a marvelous answer! How fortunate are her students. They had a rare teacher, one with a beginner's mind.

Even more recently, I learned of yet a third way to solve the problem. I gave six toothpicks with the usual instructions to a district ranger of the US Department of Agriculture Forest Service. Seated at his kitchen table, he laid the toothpicks on the table's top, looked at them for a few seconds while his young son watched; he then broke each toothpick in two.

The boy turned to me with a questioning voice and said, "He broke them."

"He didn't tell me I couldn't," replied his father as he made four equilateral triangles on the tabletop, with one piece left over.

What does breaking matches have to do with mediation? First, it demonstrates that most individuals become stuck within their self-imposed limitations from their years of social conditioning. Second, it shows that it is a rare socialized individual who has managed to retain a childlike beginner's mind.

Breaking matches is like going into mediation with *little specific knowledge* of the conflict with which you are about to deal. That way, you are as detached and unbiased as possible because you have only minimal, general information about the dispute prior to entering the arena. It is thus difficult to form opinions about who has done what or why because you do not know. You therefore remain open to possibilities—have a beginner's mind.

Because I (Chris) purposefully know little about a conflict I am going to mediate, once in the process, I have each party in turn tell me their perceptions of the dispute. The exercise is for each party to educate me about the dispute from its understanding of the whole. As each explanation unfolds, the side recounting it not only clarifies their own understanding of its perceptions but also the other party (or parties) hears for the first time the whole of the other's story from the other's point of view. During this storytelling, I learn what the dispute is about because I hear it from different sides and am thus able to find common ground, differences, negotiable areas, quagmires, and hidden potential for resolution.

With a beginner's mind, it is easier for my intuition to preside equally with my intellect and open my creative space. This open space, which allows intuition the freedom to exercise its creative powers, is perhaps the greatest value of having a beginner's mind.

Because intuition is so important to the beginner's mind, it is necessary to examine it, albeit briefly. Intuition, the knowing beyond knowledge, has been widely accepted since ancient times as the forerunner of deep inner truth and creativity. It is generally characterized as an instantaneous, direct grasping of reality, the source of our deepest truths, those of unquestionable knowing that we call "axiomatic."[15] Even John Stuart Mill, the pillar of the empirical method, stated that "the truths known by intuition are the original premises from which all others are inferred."[16]

It is well known, for example, that Niels Bohr, Albert Einstein, Sir Arthur Eddington, Eugene Wigner, and Erwin Schrödinger intuited the principles of quantum physics, where it is understood that the concepts of time, space, and the conservation of forces are based on intuitive insight, if not on faith. "Intuition," says Russian revolutionary thinker Pitirim A. Sorokin, "is more than a guide to truth; it seems also to be the ultimate foundation for our understanding of beauty and good because our aesthetic and moral judgments are based on deep subjective feelings."[17]

But, intuition has been clouded by ambiguity and controversy for the last century or so and regarded as a meaningless by-product of unconscious processes. At best, modern academia often assumes it to be lucky guesses based on gut hunches, creative flashes, and momentary insights. At worst, academic scholars, who pride themselves on discovering logical, scientific solutions to difficult problems, see intuition as irrational concoctions of the unconscious mind, based on memories, habits of thought, social conditioning, or emotional predispositions.

Intuition is therefore implied to be unreliable, unscientific, irrational, and purely subjective with no foundation in measurable reality, which is a very different view from ancient times, when it was considered "the source of deep, inner truth." It is, nevertheless, this latter meaning of intuition to which I (Chris) subscribe. I say this advisedly based on years of unerring experience.

There is yet another value to having a beginner's mind: the freedom to be authentically oneself, which the intuitive-creative edge brings out and to which people respond.

Being Oneself

When all is said and done, it is most important to just be yourself, be authentically you. Being authentically yourself has two components: authenticity and being. Authenticity is the condition or quality of being trustworthy or genuine. Beyond any dictionary definition, authenticity is the harmony between what you think, say, and do and what you really feel—the motive in the deepest recesses of your heart. You are authentic only when your motives, words, and deeds are in harmony with your attitude: freedom from guile.

As Ralph Waldo Emerson noted, "Your attitude thunders so loudly that I can't hear what you say."[18] Your attitude is the visible part of your behavior, but your motives are hidden from view. When your visible behavior is out of harmony with your motives, your attitude points to a hidden agenda.

Being is more difficult to explain to the Western mind. The best explanation of "being" that I have found is in the books by Eckert Tolle.[19]

When Zane, my wife, and I (Chris) took our morning walks in Las Vegas, Nevada, we saw spadefoot toads on the lawns, cottontail rabbits in the undeveloped lots, and occasionally a roadrunner, which is a marvelous bird in the cuckoo family. None of these animals went out of its way to please us, and yet we were thrilled each day just to see them and to wish them well. They were just "being," and in their "beingness" they brought us great joy. They did not have to do anything or be anything other than what they were doing and being.

Your mission as a mediator is simply to act as yourself and to give what you can to the best of your ability, one mediation at a time—no more and no less. To fully understand this concept, you would do well to see the movie *It's a Wonderful Life.*

In the small town of Bedford Falls, so the story goes, lives a young man, George Bailey, who cannot wait to leave his hometown to see and conquer the world. But, for one reason or another, he never leaves. Being altruistic in his outlook on life, he is other-centered and keeps passing by his chances to go to college and beyond to the benefit of others.

Finally, however, facing bankruptcy just before Christmas, through no fault of his own, George decides that he is worth more to his family dead than alive because of his life insurance policy. He therefore tries to kill himself by jumping off a bridge into the river, but an angel, Clarence Oddbody, is sent to save him. Clarence, however, cannot convince George that his life has any value. Adamant that his life is worthless, George wishes he had never been born, and Clarence grants his wish.

To the townspeople, George never existed, so while he knows everyone, no one knows him. He sees how the town would have developed and how the people would have fared had he never been born. George finally sees and understands just how many lives the ripples of his actions have affected by his just being who he is: a simple man who never left his hometown, who never conquered the world. He was perfectly himself, and that was all God asked of him: just to act as himself and to give what he could to the best of his ability, one day at a time.

So, as mediator, relax and be yourself. In turn, your example will allow participants to be themselves and to respond to what you have to offer them even as you respond to what they have to offer you, which means that you—and they—will always be learning.

The Continual Learning Curve

As a mediator, you are always in school, with much to learn and no hope of learning it all. Nevertheless, you have a full and continuing responsibility to study disciplines involved in the disputes you mediate in order to improve your awareness, skills, and abilities. Above all, it is critical to work diligently on personal, unresolved, familial matters. There is a perpetual need for personal growth in order for you, as mediator, to free yourself from the unwanted psychological and emotional baggage from your years of social conditioning.

The freer you are of your own dysfunctional baggage, the closer you are to having the all-important detachment and creativity of a beginner's mind. Along the way, you will find that you do not and cannot have all the answers. This not only is okay in and of itself but also is a blessing that perpetuates the wonder of discovery, especially for the parties with whom you are working.

Not Knowing an Answer Is Okay

Although ignorance is thought of as the lack of knowledge, there is more to it than that. Our sense of the world and our place in it is couched in terms of what we are sure we know and what we think we know. Our universities and laboratories are filled with searching minds, and our libraries are bulging with the fruits of our exploding knowledge, yet where is there an accounting of our ignorance?

Ignorance is not okay in our fast-moving world. We are chastised from the time we are infants until we die for not knowing an answer someone else thinks we *should* know. If we do not know the correct answer, we may be labeled as stupid, which is not the same as being ignorant about something. Being stupid is usually thought of as being mentally slow to grasp an idea, but being ignorant is simply not knowing the acceptable answer to a particular question.

My (Chris's) favorite answer to a question from an audience member is a purposeful "I don't know," which not only allows me to discover some heretofore hidden secret but also affirms that neither *should* be all-knowing or in charge of the universe. In my ignorance, I find the incredible freedom to accept the frailty of what it means to be human: to be simply what I am.

Society's preoccupation with building a shining tower of knowledge blinds us to the ever-present dull luster of ignorance underlying the foundation of the tower, from which all questions must arise and over which the tower of knowledge must stand. Each new brick in the tower of knowledge is born of a question that illuminates our ignorance. Yet, ignorance, which often is seen as negative, is but a point along the continuum of consciousness, as are knowledge and the knowing beyond.

The quest for knowledge in the material world is a never-ending pursuit, but the quest does not mean that a thoroughly schooled person is an educated person or that an educated person is a wise person. We are too often blinded by our ignorance of our ignorance, and our pursuit of knowledge is no guarantee of wisdom. Hence, we are prone to becoming the blind leading the blind because our overemphasis on competition in nearly everything makes looking good more important than being good. The resultant fear of being thought a fool and criticized therefore is one of the greatest enemies of true learning.

Although our ignorance is undeniably vast, it is from the vastness of this selfsame ignorance that our sense of wonder grows. But, when we do not know we are ignorant, we do not know enough to even question, let alone investigate, our ignorance.

No one can teach another person anything. All one can do with and for someone else is to facilitate learning by helping the person to discover the wonder of their ignorance. By asking an appropriate question in an appropriate way, you may be able to help that person become aware of their ignorance in a given area without stealing their dignity.

A teacher is but a "midwife," as the Greek philosopher Socrates said, because once a person realizes their ignorance and begins in earnest to search for understanding, that person slowly comes to see that such understanding can only be drawn out from within. Understanding, after all, is the unique perspective of each and every person.

Success or Failure Is the Interpretation of an Event

It is critical for a mediator to understand that success or failure is the interpretation of an event—not the event itself. The Greek philosopher Epictetus hit the mark when he wrote, "Men are disturbed not by things, but by the view which they take of them."[20]

Success and failure are apparent opposites of the same dynamic. They are merely degrees of magnitude along a continuum. The interpretation may be somewhat influenced, however, by who does the interpreting. Knowing what you want to get out of an event, you may interpret the outcome as a huge success, whereas others, whose perceptions of success demand a different outcome, may interpret the same event as a dismal failure.

How, then, do you measure success in mediation when you may not see a tangible outcome of the process? Were you to either hope for or expect a certain outcome, you may not see anything that even closely resembles it. By seeking a certain outcome, however, you would have an attachment to the result and therefore bias it. So, how do you succeed? You succeed by having no expectations and letting the process guide itself to the necessary conclusion.

The labels of success or failure are usually assigned to the perceived outcome or product of an event rather than to the actual process of learning embodied in the event itself. Therefore, when you abdicate your right to interpret the event, such as a mediation process, success or failure becomes a judgment determined by others, which may be termed the *cultural trance* because the value of an individual's spontaneous creativity usually is smothered by the prevailing social standards.

The cultural trance is the uncritical acceptance of every *should, ought,* and *must* thrust on us from childhood onward by myriad external sources. If you look to *others* for your self-esteem, success involves a visible accomplishment of which others must approve. Lack of such visible accomplishment, therefore, is deemed a failure. Success means you do it yourself, and failure means you need help. Because success is usually measured in terms of how much money or power one has, failure is simply a measure of perceived lack. Success is a measure of being in control of circumstances, and failure is a measure of not being in control of circumstances.

Another measure of success or failure in life, however, is whether you are doing what you are passionate about. Mythologist Joseph Campbell called it "following your bliss." If you are doing what you really want to do, regardless of material returns, then you have a success of the heart, no matter what anyone else thinks. Others can measure only the appearance of success based on *their* definition. Yet, an apparent failure in the short term can prove a success in the long term. Failure can then be viewed as delayed or postponed success.

If, therefore, you allow others to define you, you allow them to determine your success or failure. If you accept failure on this basis, then the point at which you acquiesce is the point at which you are defeated.

One of my (Chris's) favorite stories is about Babe Ruth, the baseball player. Ruth had struck out two or three times in one game. When the game was over, a sports reporter asked him what he was thinking about as he struck out.

"Hitting a home run," he said.

"Well," asked the reporter, "how does it feel to fail?"

"I didn't fail," replied Ruth. "Every time I get a strike, I'm one swing closer to the next home run, and you know who has to worry about that!"

Babe Ruth had learned two very important lessons. The first was that to be a winner, he had to be willing to risk being a loser, and the degree to which he was willing to risk losing determined the degree to which he was ultimately capable of winning. Ruth was the home run king only because he was first willing to be the strikeout king. We (in the generic) all have to be willing to be poor at something for a while so that, through practice, we can excel.

The second lesson was to be consistent in his efforts, which Winston Churchill put well when he said, "Success is going from failure to failure with enthusiasm."[21] The crowd in the bleachers did not bother Ruth. He smiled at them in the same, quiet way when they booed him as he did when they cheered him. Babe Ruth knew who and what he was, a champ. Why? Because he was what the Buddhists call "one pointed."

Being one pointed means that you are totally focused on what you are doing; it is therefore necessary to divest yourself from all diversions. To this end, you would be wise to sign and honor both a conflict of interest statement and a confidentiality statement pertaining to each and every dispute resolution that you facilitate.

Assisting Parties in Clarifying and Resolving Their Conflict

If, in the process of mediation, you find nothing good to say about a co-mediator—remain silent. Discussion between and among mediators concerning any given case, but particularly an active one, is to be conducted solely in private. It is also imperative that you enter into a conflict being handled by another mediator or other mediators only after fully conferring with them and only after receiving everyone's approval, which includes all persons involved in the dispute.

In addition, to assist parties in clarifying and resolving their conflict, you must understand the dynamics of change in both the ecological sense and the human dimension. This means you must be able to help the people understand and cope with change, as a creative process to be accepted, lived with, and adapted to as an opportunity, rather than resisting change as a condition to be avoided. Change, after all, is a universal constant, and its effects, while unavoidable, always novel, and irreversible, can be purposefully guided to some extent.

You must also be well versed in both functional and dysfunctional coping mechanisms, homeostasis, and boundaries through having done personal work with your own unresolved familial issues. Such understanding is critical because how one copes with change is, in large measure, predetermined by childhood learning experiences in the form of social conditioning.

When Potential Resolution Is in Violation of Public Policy or Law

At times, you may find yourself in a situation where the parties moving toward an agreement are doing so contrary to public policy or in violation of the law. Because you have no right to impose standards of behavior on the parties, it may be ethically necessary for you to withdraw from the process. This is an important consideration because you are both legally and ethically without authority to enforce law or to act in any way as an agent of investigation or law enforcement. As soon as such a situation arises, you must consult with the proper authority, who in turn will recommend the action to be taken.

Discussion Questions

1. Why is leadership, and therefore mediation, the art of being a servant?

2. How is the role of a conflict mediator simultaneously one of being a "teacher"?

3. Why is it important to maintain (= protect) each person's dignity?

4. What is the fallacy of trying to rescue disputants?

5. What does it mean to be a sieve, as opposed to a sponge?

6. What does it mean to have a "beginner's mind," and why is it important in resolving conflicts?

7. Is it acceptable *not* to know the answer to a question?

8. How would you deal with the potential resolution of a conflict when said resolution would be in violation of public policy or law?

9. Is there a particular question you would like to ask?

Endnotes

1. Gerald Corey. *Theory and Practice of Counseling and Psychotherapy*. 3rd ed. Brooks/Cole, Monterey, CA, 1986.
2. Quotations Book. Francis Bacon. http://quotationsbook.com/quote/1246/ (accessed March 21, 2010).
3. James Allen. *As a Man Thinketh*. Grosset & Dunlap, New York, 1981.
4. The foregoing five paragraphs are based on Chris Maser. *Decision Making for A Sustainable Environment: A Systemic Approach*. CRC Press, Boca Raton, FL, 2013. 304 pp.
5. ThinkExist.com. Henry Ford Quotes. http://thinkexist.com/quotation/coming_together_is_a_beginning-keeping_together/146314.html (accessed March 21, 2010).
6. Stephen Mitchell. *Tao Te Ching*. Harper Perennial, New York, 1992.
7. Hans Bleiker. http://www.ipmp.com/about/ (accessed December 1, 2010).
8. Ibid.
9. Kenneth W. Estes and Robert Debs Heinl. *Handbook for Marine NCOs*. 5th ed. Naval Institute Press, Annapolis, MD, 1996. 384 pp.
10. Henry Wadsworth Longfellow. sxhttp://lhmci.com/study/qualities/sensitivity.html (accessed March 21, 2010).
11. Abraham Maslow. *Toward a Psychology of Being*. 2nd ed. Van Nostrand, Princeton, NJ, 1968. 240 pp.
12. Ibid.
13. John F. Kennedy. *Profiles in Courage*. Harper & Row, New York, 1961. 266 pp.
14. Louis Fischer. *The Essential Gandhi: An Anthology of His Writings on His Life, Work, and Ideas*. Random House, New York, 2002. 368 pp.
15. Jeffrey Mishlove. Intuition, the Source of True Knowing. *Noetic Sciences Review*, 29(1994):31–36.
16. John Stuart Mill. Introduction. In *A System of Logic*. Longman, London, 1884. (Original work published 1843.) http://www.marxists.org/reference/archive/mill-john-stuart/1843/logic.htm (accessed March 21, 2010).

17. Pitirim A. Sorokin. Excerpted from *The Crisis of Our Age*. http://www.intuition. org/sorokin.htm (accessed March 21, 2010).
18. Famous Quotes & Authors. Ralph Waldo Emerson Quotes and Quotations. http://www.famousquotesandauthors.com/authors/ralph_waldo_emerson_ quotes.html (accessed March 21, 2010).
19. (1) Eckert Tolle. *The Power of Now*. New World Library, Novato, CA, 2004. 193 pp.; (2) Eckert Tolle. *A New Earth*. Plume, New York, 2006. 313 pp.
20. BrainyQuote. Epictetus Quotes. https://www.brainyquote.com/quotes/ quotes/e/epictetus149127.html (accessed June 26, 2017).
21. Winston Churchill. http://fbicgirls.com/index.php?option=com_content& view=article&id=89:the-courage-to-fail&catid=16:blog&Itemid=2 (accessed March 22, 2010).

Section II

The Legacy of Resolving Environmental Conflicts

9

Practicing Mediation of Conscience

Good conflict resolution is a meticulous practice in democracy. It is thus critical to understand something about democracy as a practical concept. Democracy is a system of shared power with checks and balances, a system in which individuals can affect the outcome of political decisions. People practice democracy by managing social processes themselves. Democracy is another word for self-directed social evolution.

Democracy in the United States is built on the concept of inner truth, which in practice is a tenuous balance between spirituality and materialism. One such truth is the notion of human equality, in which all people are pledged to defend the rights of each person, and each person is pledged to defend the rights of all people. In practice, however, the whole endeavors to protect the rights of the individuals, while the individuals are pledged to obey the *will* of the majority, which may or may not be just to each person.

The "will" of the majority brings up the notion of freedom in democracy. Nothing in the universe is totally free because everything is in a relationship with everything else, which exacts various degrees of constraints. Just as there is no such thing as a truly "free market" or an "independent variable," so individual and social autonomy are protected by moral limits placed on the freedom with which individuals and society can act. To this effect, author Anna Lemkow listed four propositions of freedom: "(1) An individual must win freedom of will by self-effort, (2) freedom is inseparable from necessity or inner order, (3) freedom always involves a sense of unity with others beyond differences, (4) freedom is inseparable from truth—or, put the other way around, truth serves to make us free."[1]

Lemkow went on to say the following:

> We tend to think of freedom as dependent on circumstantial or external factors, but these propositions point us inward, suggesting rather that freedom is a state of consciousness and ... depends on ourselves. Indeed, it is something to be won, something to be attained commensurately with becoming more truthful, or more attuned to and aligned with the abiding inner, metaphysical, or moral order or law.
>
> Socio-political and economic freedom or liberty, in turn, would depend (at least in the longer term) on the predominant level of consciousness of the citizenry.[2]

Lemkow was positing that a human being is not completely free to begin with, but possesses the potential capability of self-transformation in the direction of fuller freedom. Beyond this, democracy requires respect for others and excitement in the exchange of ideas. People must learn to listen to one another's ideas, not as points of debate but as different and valid experiences in a collective reality. While they must learn to agree to disagree at times, they must also learn to accept, like blind people feeling the different parts of an elephant, that each person is initially limited by their own perspective. When these things happen, people are engaged in the most fundamental aspects of democracy and come to conclusions and make decisions through participative talking, listening, understanding, compromising, and agreeing.[3]

In a democracy, *connection* and *sharing* are central to its viability because a democracy only works when it is being practiced. People are not required to separate feelings from thoughts concerning a topic. Their roles—as teachers, students, leaders, mediators, and followers—fluctuate within and across the issues. The importance of a democratic system lies in its connection to people's lives, their own experiences, and the real problems and issues they face daily. Practicing democracy can be thought of as education in and for life. And, because people have within them the seeds of greatness, as teacher Myles Horton said, "It is not a matter of trying to fill people up but rather of fulfilling them."[4]

The challenge, therefore, is to engage people in the democratic process, which is difficult when they confuse the government with the administration of the government or when the administration becomes so dysfunctional that people despair in their seeming inability to fix it. But then, a perfect government does not exist.

A just government must be founded on truth, not knowledge—something we in the United States have all too swiftly forgotten. To achieve such government, it must be based on service (where people are other-serving) rather than power (where people are self-serving). Therefore, environmental protection is only possible if the government is accountable to its people beyond special interest groups and political lobbyists.

We live in an increasingly complex society of intense competition and materialism. In such a society, people commit foul acts in order to gain power, both through positions of authority and financial maneuvering. People commit such acts while falsely expecting to benefit by them, thinking that such benefits will somehow bring happiness. But, in the end, as Socrates warned, the guilt of the soul outweighs the supposed material gains. Thus, because people lack perfect knowledge and perfect motives, democracy must be continually practiced and continually improved through that practice.

Nevertheless, the *people are the government*, but they can govern only as long as they elect to use the constitutional system for empowerment. It is important to understand that empowerment is self-motivation. No one can empower anyone else; one can only empower oneself. One can, however, give others the psychological space, permission, and skills necessary to empower themselves and then support their empowerment. Beyond that, one can help in the process of empowerment and can increase the chances of success by recognizing another's accomplishments each step of the way. That is true democracy.

Therefore, in the case of the United States, where the government is of the people, by the people, and for the people, when the people empower themselves, they are the government, and it is the administration of that government that resides in Washington, D.C., not the government itself. The administration becomes the government only when the people turn their power over to the administration and in effect say, I'm a victim and can't change the system, or take care of me.

Democracy is a viable system because, by inviting a constant reinterpretation of itself, it rests on a self-reflective principle of always being in a state of becoming, which continually interweaves it within the intimacy of life. For democracy to remain viable, it must be used because it is an interconnected, interactive system of balancing and integrating contrasting perceptions of data, facts, and truth. A working democracy is thus predicated on finding the point of balance through compromise in such a way that the rifts between opposites can be minimized and healed.

Compromise and the Point of Balance

Mediation is the way to compromise and the point of balance that resolves conflicts. The *mandorla*, a symbol of unity, is a prototype of conflict resolution that has long been secreted in the gathering dust of medieval Christianity.

A mandorla is two overlapping circles with an almond-shaped area in the middle, where the contents of each circle integrate. When thus put together, their areas of overlap and integration—their common root—can be found, and perceived opposites can be balanced. And, all opposites have a common root because they are, after all, merely different perceptions of the same reality. For example, a glass of water is half full or half empty, depending on one's point of view, but the level of water is the same in either case.

Once the overlap is identified, acknowledged, and accepted, people can begin working collectively, extending the area of overlap and integration. Although the overlap is tiny at first, like the sliver of a new moon, it is a beginning, the first healing of the split between opposites. With diligent work and the passage of time, the sliver becomes as a quarter moon, then a half moon and a three-quarter moon, until that point is reached where the two circles become as one: a full moon, unity, total healing.

The mandorla, as a symbol, a process, and a metaphor, fits every social-environmental problem imaginable. The mandorla thus seems a logical metaphor of social-environmental sustainability, which is the necessary next stage of social evolution—the ultimate expression of a working democracy—if society, as we know it, is to survive. If, however, we are to map the country of the mandorla to our best advantage, we must treat one another with compassion and justice while we explore the hidden potential of the almond-shaped land of overlap and integration.

A Curriculum of Compassion and Justice

For mediation to be successful, the process must be as gentle and dignified as possible, which means that it must be a continual lesson in compassion and justice taught through the mediator's example. All parties must emerge with their dignity intact if anything is to be resolved. It is therefore important to remember that *now* is always the time for compassion and justice, because, as Mahatma Gandhi pointed out, "An eye for an eye only makes the whole world blind."[5] In this sense, mediation, as a democratic process, is perhaps at its best when the people involved must continue dealing with one another after the dispute is resolved.

Compassion is the deep feeling of sharing another's suffering, of giving aid and support to another person in their time of need, which is the act of forgiving another's perceived trespass—of extending mercy. The essence of compassion is best acknowledged in a French proverb: "To know all is to forgive all." This says that, as I do the best I can in all I do, so does everyone else, so where, therefore, is the judgment? Thus, when we forgive all, when we fix no judgment and place no blame, we have compassion.

When someone is unkind to me (Chris), I do my best to accept that it is not a personal act, but rather one that reflects the other person's inner distress, and I know that neither of us will gain anything if I shame or shun that person. I do not always succeed in my endeavors, however, and I cannot count the times I have fallen short of my ideal of unconditional compassion, which to me is the understanding through which the act of forgiveness may flow.

Forgiveness, in turn, is to see the fear and the pain out of which another person acts and to extend love as an alternative. To have compassion therefore demands far greater courage than does retaliation in any form at any time because compassion demands that one is responsible for one's own behavior and thereby abnegates the role of victim.

Although I (Chris) cannot experience how the other person is feeling, I can ask. If that person can relax long enough to answer, I may be able to imagine myself for a moment in their situation and see if I would act any differently under the particular circumstances. I have inevitably found that I would do the same if I were in the other person's shoes. It is therefore up to me to forgive rather than up to the other person to change, which means that I must do unto others as I would have them do unto me in any given circumstance.

We have often heard it said that "all's fair in love and war" and that "nothing's fair in life." If this is the way people really feel, perhaps we must shift our attention from fairness to justice or that which is just. If I (in the generic) am just, I am honorable, consistent with the highest morality, and equitable in my dealings and actions. If I do unto others as I would have others do unto me, then I can be a just person.

One can also choose to be *just* in a practical sense. Being just and equitable with others encourages them to treat you justly, which is another lesson you can teach by example for it is wise to cultivate such kindly behavior for yourself by extending it first to others. And, if for some reason the person with whom you were just is unjust with you, they give you an opportunity to practice acceptance, detachment from personalization, and compassion. Being compassionate requires nothing from anyone else, only your courage and the wisdom to love enough to forgive. Such compassion is one of the gifts passed forward through an ideal mediation process.

Mediation as a Gift Is Free, But as a Trade Has a Cost

What do a gift and a trade have in common with mediation? They have much to do with it because mediation must be an unconditional gift. To define an unconditional gift, consider what you, as a mediator, might expect out of the process. Do you want something specific to happen, a certain outcome, for example, or is any outcome okay?

A gift is free of expectations, but a trade has a specific outcome attached to it. Both a gift and a trade are circumstances to which the recipient must respond, and the choice of response is one of either trust or distrust.

A gift is a feeling made visible through an object, but the object is not the real gift. The feeling is the real gift because it is the cause for giving the object.

A gift, which most people think of as an object, is free of conditions. If I (in the generic) give you a true gift, I have, by definition, also given you title and ownership free of encumbrances. You may do with it as you wish because I have no vested interest in what you do with what you own, which means no strings are attached. Further, your unconditional acceptance of my gift is your unconditional gift to me because I cannot give you my gift if you refuse to accept it. This thought is in keeping with an indigenous American proverb: "You must humble yourself to receive before you can truly give."

If, for example, I give you a chocolate cake, but you are on a diet so you graciously accept the cake and give it to someone else, then you have accepted my gift and given me one through your acceptance. In addition, you have given a gift of your own by passing on the cake. If the second person accepts the cake but does not like chocolate and gives the cake to a third person, then the second person has accepted your gift and given one of their own. In each case, the apparent or perceived gift was the cake, but the true gift was the unconditional love embodied in the thought of giving by one person and in the thought of receiving by the other. At no time was the real gift the cake itself.

A trade, on the other hand, is an exchange of one thing for another, beginning with the thought or expectation of a specific outcome. By definition, therefore, a trade cannot be a gift because a gift, truly given and received, is something bestowed without thought of compensation. A trade, however, is a way of realizing a known expectation, a way to control the outcome.

Let us go back to the chocolate cake. If I give you a chocolate cake on the condition that you eat it, I have given you a condition—a prison cell—instead of a gift. I have traded a chocolate cake to you for your compliance with my expectations of your behavior: to eat the cake in spite of your diet. I have used the cake to control you. I have covertly said, "If you really love me, you will eat the cake for me in spite of your diet." The inference is that if you choose to honor your diet and not eat the cake, then you do not really love me. I have laid guilt on you to control your behavior and give me the outcome I want.

Unless all parties agree to the rules of the trade beforehand, the "trade" is really coercion. Such was the case with the meeting on the livestock grazing fee program discussed previously. The facilitators wanted something from the participants. They wanted compliance with their wishes, which, if forthcoming, would please the interior secretary and might bring in more business. Although the participants were not privy to what the facilitators wanted, it became clear during the mediation process that they had a hidden agenda.

People trade because an unexpected outcome will force them out of their comfort zone and make them deal with the unknown, and that is not what they want. It is the unknown—that which they think is unknowable, that which demands the risk of uncontrollable change—that is frightening.

One of the many places fear of the unknown can be seen in people is in airports. Passengers have their tickets, and their schedules are confirmed, but not beyond change. A flight is suddenly canceled or baggage is lost, and carefully controlled plans are nowhere to be found. This is distressing because the known expectation suddenly evaporates, and the worst-case scenario of the unknown becomes reality. The dreaded unknown—not knowing what to expect—has happened. You see no immediate clear choice. You have suddenly lost control.

This is when frightened people become angry and yell at ticket agents. In so doing, they are trying to trade their anger for compliance with their wishes or more comfortable boundaries within which they can operate "normally." The more frightened they are, the nastier their behavior becomes.

But, unknown expectations are also gifts of adventure, which always hold lessons about our lives if we will but look for them. One of my (Chris's) college professors, for example, had such an unexpected experience. He had purchased an expensive microscope in Switzerland, only to have it stolen in Naples, Italy, on his way home. He went to the police station to report the theft. While he stood there waiting, the police chief talked to an American sailor whose wristwatch had been stolen:

> "Which pier is your ship docked at?" asked the chief.
> "Pier 10," said the sailor.
> "I think I know who has your watch. You two come with me," said the chief motioning to both the sailor and the professor.

When they reached the pier, the chief went up to a young boy and said, "Give this man his wristwatch." The boy reluctantly complied.

This, however, did nothing for the professor's microscope. In fact, the chief told him that his problem was much more difficult to solve, probably impossible, and because the microscope was not insured against theft, the professor was probably out of luck.

The professor returned home, angry that he had been ripped off. It had, retroactively, spoiled his whole trip; it was all he could think about. He stewed about the microscope and the money for 6 months. With time, he finally accepted what was; he let go of what he could not control (the circumstance) and took responsibility for what he could control: his attitude, his response to the circumstance.

Several months later, the professor got a telephone call from customs at the Seattle-Tacoma Airport in Washington State advising him that a package had arrived for him. Lo and behold, his microscope had arrived in perfect condition. All of his stewing and internal discord had been a total waste of time and energy.

Thus, you must be detached from the outcome of your mediation and must become dispensable to all parties, as soon as possible, by helping them, should they so choose, to establish the processes necessary to resolve their own conflicts and create a vision for their own future. The gift you have to give is helping parties get ready for the next step: bringing the conflict to resolution through a shared vision on which to act.

Discussion Questions

1. How does one find the point of balance in a compromise?
2. What, in conflict resolution, has compassion got to do with justice?
3. What is the real difference between a gift and a trade?
4. Is there a particular question you would like to ask?

Endnotes

1. Anna F. Lemkow. Our Common Journey toward Freedom. *The Quest,* 7(1994):55–63.
2. Ibid.
3. Chris Maser. Vision and Leadership in Sustainable Development. Lewis Publishers, New York. 1999. 235 pp.
4. Myles Horton and Paulo Freire. *We Make the Road by Walking* (edited by Brenda Bell, John Gaventa, and John Peters). Temple University Press, Philadelphia, 1990. 256 pp.
5. Louis Fischer. *The Essential Gandhi: An Anthology of His Writings on His Life, Work, and Ideas.* Random House, New York, 1962. 368 pp.

10

Resolution: Destructive Conflict Transformed into a Shared Vision

An ancient custom of the indigenous Americans was to call a council fire when decisions affecting the whole tribe or nation needed to be made. To sit in council as a representative of the people was an honor that had to be earned through many years of truthfulness, bravery, compassion, sharing, listening, justice, and being a discreet counselor. These qualities were necessary because a council fire, by its very nature, was a time to examine every point of view and explore every possibility of a situation that would in some way affect the community's destiny.

When someone called a council, that person had to have the courage to accept the council's decision with grace because, when the good of the whole was placed before the good of the few, all were assured a measure of abundance. The timeless teaching of the council fire is that until all of the people are doing well, none of the people are doing well.

The council fire worked well for the indigenous Americans because they knew who they were culturally, and they had a sense of place within their environment. Today, however, the global society is in transition, which robs many people of their original sense of place and substitutes some vague idea of location.

This transition is largely the result of massive human migration over the last three centuries. These shifts have altered the composition of peoples and their cultural structures throughout the world. All of this activity results in growing interconnectedness, interactivity, interdependence, and cultural uncertainty as some political lines change physically and others blur culturally. Cultural uncertainty is particularly true for those people caught between two worlds, such as the warring religious factions around the world, where millions of refugees have not only their sense of culture disrupted but also their sense of place transformed into an alien location, where their lives hang in limbo. These changes pose necessary questions for some people (such as, Who are we as a culture?), which must be answered before a statement of vision and goals can be fruitfully considered.

Who Are We as a Culture?

Who are we now, today? This is a difficult, but necessary, question for people to deal with if they want to create a vision for the future. They must decide, based on how they define their cultural identity, what kind of vision to create. The self-held concept of who a people are is critical to their cultural future because their cultural self-image will determine what their community will become socially, which in turn will determine not only what their children will become socially but also how they will affect their natural environment.

Thus, how well a people's core values are encompassed in a vision depends first on how well the people understand themselves as a culture and second on how well that understanding is reflected on paper. Let us consider three examples: the Japanese, indigenous Canadians, and indigenous Americans.

In Japan, a religious system of belief, today known as Shinto, has been observed since the founding of the country. Shinto gained systematic form spontaneously from within the social life of communities. As a result, it has no specific founder or clearly defined body of scripture. Since ancient times, the Japanese have transmitted the legends and myths of the deities or *kami* as a genealogy of their way of life.

Shinto, in its broadest sense, refers to the entirety of indigenous culture, as opposed to Buddhism and other religious systems imported from outside Japan. Shinto is established against a background of hydraulic rice agriculture, which is uniquely suited to Japan's warm and humid climate.

When used in the narrow sense, Shinto refers to the rites offered to deities, primarily those deities of heaven and Earth listed in classical Japanese works of the ancient period. The physical facility used for the performance of this worship is called *jinju* or "shrine."

That Nature and natural phenomena are revered as deities is a result of the Japanese view of Nature as a kind of parent that nurtures life and provides limitless blessings. Shinto shrines all over Japan are surrounded by luxuriant groves of trees. Backed by the Shinto view that untouched Nature is itself sacred, the groves surrounding the shrines are themselves an important composite element of each shrine.

About 1,300 years ago, Emperor Tenmu ordained the practice of removing the old shrine, such as the Grand Shrine of Ise in Ise City, every 20 years and rebuilding a new exact replica next to it. Why Emperor Tenmu stipulated that the rebuilding of the shrine should take place every 20 years is not clearly known, but it is most likely that 20 years was considered to be the optimum period for allowing the exact replication of the Grand Shrine, considering that it has a thatched roof, has unpainted or otherwise unpreserved structures, and is erected on posts sunk into the ground with only the benefit of foundation stones.

Twenty years is perhaps also the most logical interval in terms of passing from one generation to the next the technological expertise needed for the exacting task of duplicating a shrine. The Shinto shrine can be thought of as sacred architecture created from within the prayer and technical skills of the Japanese people themselves. Passing technical skills and the prayer embodied in the sacred architecture from generation to generation is the context within which lies the real significance of the regular rebuilding. The cultural knowledge has thus far been passed on for 1,300 years without change.

Herein lies the challenge for Japan. Although the sacred Shinto architecture can be passed from generation to generation through the perpetuation of the shrines, the belief system is being eroded by the introduction of Western philosophy. This psychic split seems to occur largely because Shinto is based in the belief of Nature's primacy through cycles and processes. In contrast, might Western philosophy be described as embedded in the primacy of linear technology and consumerism? The Japanese are therefore caught between two worlds: their ancient, ritualized, spiritual world and the new Western, materialistic "free"-market one.

The indigenous Canadians have a somewhat similar dilemma. They have departed from their old culture because they have—against their will—been forced to adopt European-Canadian ways, which means they have given up or lost ancestral ones. Yet, they have not—by choice—totally adopted white culture and want to retain some degree of their ancestral culture. Thus, the question they must ask and answer is, Which of our ancestral ways still have sufficient cultural value for us to keep them? Which of the white ways are we willing to adopt? How do we put the chosen elements of both cultures together in such a way that we can today define who we are culturally?

Although indigenous Americans have a similar task as the indigenous Canadians, they have even fewer options because the European-Americans were more culturally destructive, forcing them to adopt foreign ways. Nevertheless, the question of who we are today is still valid for any vision the people of an Indian reservation in the United States might want to create for their own lands. Here, I (Chris) suggest, we all face the question of "who are we today as a culture," especially at the community level. This is an important question because how it is answered will determine the legacy inherited by our children.

For example, in 1993, I was asked to review an ecological brief for a First Nation of indigenous Canadians whose reservation was located between the sea and land immediately upslope from the reservation that a timber company wanted to log. The problem lay in the fact that the timber company could only reach the timber it wanted to cut by obtaining an easement through the reservation, which gave the First Nation some control over the timber company. The First Nation, on the other hand, wanted some control over how the timber company would log the upper-slope forest because the outcome would, for many years, affect the reservation, which is just below the area to be cut.

Before meeting with the timber company, the First Nation's chief asked me for some counsel. My reply was as follows:

> Before I discuss the ecological brief I've been asked to review, there are three points that must be taken into account if what I say is to have any value to the First Nation. What I'm about to say may be difficult to hear, but I say it with the utmost respect.

Point 1: Who are you, the First Nation, in a cultural sense? You are not your old culture because you have—against your will—been forced to adopt some white ways, which means you have given up or lost ancestral ways. You are not—by choice—white, so you may wish to retain some of your ancestral ways. The questions you must ask and answer are (1) What of our ancestral ways still have sufficient value that we want to keep them? (2) What of the white ways do we want to, or are we willing, to adopt? (3) How do we put the chosen elements of both cultures together in such a way that we can today define who we are as a culture?

Point 2: What do you want your children to have as a legacy from your decisions and your negotiations with the timber company? Whatever you decide is what you are committing your children, their children, and their children's children to pay for the effects of your decisions unto the seventh generation and beyond. This, of course, is solely your choice, and that is as it should be. I make no judgments. But whatever you choose will partly answer Point 3.

Point 3: What do you want your reservation to look like and act like during and after logging by the timber company? How you define yourselves culturally, what choices you make for your children, and the conscious decisions you make about the condition of your land will determine what you end up with. In all of these things, the choice is yours. The consequences belong to both you and your children and your children's children into the unforeseeable future.

After they answered these questions for themselves, they had to determine what they wanted to leave for their children.

What Legacy Do We Want to Leave Our Children?

Once a group of people, whether a community, such as an Indian tribe or perhaps your own hometown, has defined itself culturally, it can then decide what legacy it wants to leave its children. This must be done consciously because whatever decisions the group makes under its new cultural identity, the consequences of those decisions are what the group is committing its children, their children, and their children's children to pay.

Having defined who they are culturally, and having determined what legacy they want to leave their children, the people of a community are now ready to craft a vision of what they want because only now do they really know.

The rest of my reply to the First Nation applies here:

Now to my comments: This is a difficult task at best. As with every definition, it is a human invention and has no meaning to Nature. Therefore, you must tell the timber company, clearly and concisely, what the terms in this ecological brief mean to you and how you interpret them with respect to the company's actions that will affect your reservation.

First: Every ecosystem functions fully within the limits imposed on it by Nature and/or humans. Therefore, it is the type, scale, and duration of the alterations to the system—the imposed limits—that you need to be concerned with. If your reservation looks the way you want it to and functions the way you want it to, then the question becomes, How must we and the timber company behave to keep it looking and functioning the way it is? If, on the other hand, your reservation does *not* look the way you want it to and does *not* function the way you want it to, then the question becomes, How must we and the timber company behave to make it look and function the way we want it to? But regardless of your decisions or the company's actions, your reservation will always function to its greatest capacity under the circumstances Nature, you, and the company imposes on it. The point is that your decisions and the company's actions, excluding what Nature may do, will determine how your reservation both looks and functions. This reflects the importance of the preceding Point 3 and what you decide.

Second: If you want the landscape of your reservation to look and function in a certain way, then how must the timber company's landscape look and function to help make your reservation be what you want it to be? Keep in mind that the landscape of your reservation *and* that of the company are *both* made up of the collective performance of individual stands of trees or "habitat patches." Therefore, how the stands look and function will determine how the collective landscape looks and functions.

Third: Remember that any undesirable ecological effects are also undesirable economic effects over time. Your interest in your reservation will be there for many, many years, generations perhaps, but the company's interest in the forest may well disappear just as soon as the trees are cut. So, the company's short-term economic decision may be good for them but may, at the same time, be a bad long-term ecological and a bad long-term economic decision for you.

Fourth: To maintain ecological functions means that you must maintain the characteristics of the ecosystem in such a way that its processes are sustainable. The characteristics you must be concerned about are (1) composition, (2) structure, (3) function, and (4) Nature's disturbance regimes.

The composition or kinds of plants and their age classes within a plant community create a certain structure that is characteristic of the plant community. It is the structure of the plant community that in turn creates and maintains certain functions. In addition, it is the composition, structure, and function of a plant community that determines which animals can live there and how many. If you change the composition, you change the structure, you change the function, and you affect the animals. People and Nature are continually changing a community's structure by altering its composition, which in turn affects how it functions.

For example, the timber company wants to change the forest's structure by cutting the trees, which in turn will change the plant community's composition, which in turn will change how the community functions, which in turn will change the kinds and numbers of animals that can live there. These are the key elements with which you must be concerned because an effect on one area can—and usually does—affect the entire landscape.

Composition, structure, and function go together to create and maintain ecological processes both in time and across space, and it is the health of the processes that in the end creates the forest. Your forest is a living organism, not just a collection of trees—as the timber industry usually thinks of it.

Fifth: Scale is an often-forgotten component of healthy forests and landscapes. The treatment of every stand of timber is critically important to the health of the whole landscape, which is a collection of the interrelated stands.

Thus, when you deal only with a stand, what is ignored is the relationship of that particular stand to other stands, to the rest of the drainage, and to the landscape. It's like a jigsaw puzzle, where each piece is a stand. The relationship of certain pieces (stands) constitutes a picture of the drainage. The relationship of the pictures (drainages) makes a whole puzzle (landscape). Thus, relationships of all the stands within a particular area make a drainage, and the relationships of all the drainages within a particular area make the landscape.

If one piece is left out of the puzzle, it is not complete. If one critical piece is missing, it may be very difficult to figure out what the picture is. So, each piece (stand) is critically important in its relationship to the completion of the whole puzzle (landscape). Therefore, the way each stand is defined and treated by the timber company is critically important to how the landscape, encompassing both the company's land and your reservation, looks and functions over time.

Sixth: Degrading an ecosystem is a human concept based on human values and has nothing to do with Nature. Nature places no value on anything. Everything just is, and in being it is perfect. Therefore, if something in Nature changes, it simply changes, but no value is either added or subtracted. On the other hand, whether or not your reservation becomes degraded depends on what you want it to be like, what value or values you have placed on its being in a certain condition, to produce certain things for you. If your desired condition *is* negatively affected by the company's actions, then your reservation *becomes* degraded. If your desired condition is *not* negatively affected by the company's actions, then your reservation is *not* degraded. Remember, your own actions can also degrade your reservation.

Seventh: It is important that you know—as clearly as possible—what the definitions in this brief really mean to you and your choices for your children and your reservation. Only when you fully understand what these definitions mean to you can you negotiate successfully with the timber company.

To negotiate with the timber company, however, the First Nation must have a vision, goals, and objectives for their reservation.

Vision, Goals, and Objectives

Although the word *vision* is variously construed, it is used here as a shared view of the future—a view based on the coalescence of myriad little, personal decisions, which in concert evolve into a big, collectively shared decision, *the vision*. Defining a vision and committing it to paper goes against our training, however, because it must also be stated as a positive in the positive, which is something we are not used to doing. Stating a positive *in the positive* means stating what we mean directly. For example, a town has an urban growth boundary that it wants to keep within certain limits, which can be stated in one of two ways: (1) We do not want our urban growth boundary to look like that of our neighbor, which is a negative stated as a positive; or (2) we want our urban growth boundary to remain within a half mile from where it is now situated, which is a positive stated as a positive.

Further, to save our planet and human society as we know it, we must be willing to risk changing our thinking in order to have a wider perception of the world and its possibilities and to validate one another's points of view or frames of reference. The world can be perceived with greater clarity when it is observed concurrently from many points of view, like the previously discussed compound eye of an insect, which can see over a hundred points of view simultaneously. Such conception requires open-mindedness in a collaborative process of intellectual and emotional exploration of what is and what could be, the results of which are a shared vision.

To illustrate the potential a group may have for committing to a shared vision, consider the movie *Spartacus*, which depicts the story of a Roman slave forced to become a gladiator and who led an army of slaves in an uprising in 71 BCE. These slaves defeated the Roman legions twice but were finally conquered by General Marcus Licinius Crassus after a long siege and battle in which they were surrounded by and had to fight three Roman legions simultaneously.

The battle over, Crassus faces the thousand survivors seated on the ground as an officer shouts, "I bring a message from your master, Marcus Licinius Crassus, Commander of Italy. By command of his most merciful Excellency, your lives are to be spared. Slaves you were, and slaves you remain. But the terrible penalty of crucifixion has been set aside on the single condition that you identify the body or the living person of the slave called Spartacus."

After a long pause, Spartacus stands up to identify himself. Before he can speak, however, Antoninus leaps to his feet and yells, "I am Spartacus!" Immediately thereafter, another man stands and yells, "No, I'm Spartacus!" Then another leaps to his feet and yells, "I'm Spartacus!" Within minutes, the whole slave army is on its feet, each man yelling, "I'm Spartacus!"

Each man, by standing, was potentially committing himself to death by crucifixion. Yet, their loyalty to Spartacus, their leader, was superseded only by their loyalty to the vision of themselves as free men, the vision that

Spartacus had inspired. The vision was so compelling that, having once tasted freedom, they willingly chose death over submitting to slavery. By withholding their obedience from Crassus, they remained free because slavery requires the oppressed to submit their obedience to the oppressor.

In more recent times, a vision of freedom and equality inspired 13 colonies to formally declare their independence from England on July 4, 1776. The vision of human freedom and equality was so strong that a whole nation, the United States of America, was founded on it. In 1836, the fall of the Alamo, the Franciscan mission in San Antonio, Texas, and the slaughter of the men defending it inspired Texans in their vision of freedom from Mexican rule. In both cases, the strength of the vision carried a people to victory against overwhelming odds.

Although a vision may begin as an intellectual idea, at some point it becomes enshrined in one's heart as a palpable force that defies explanation. It then becomes impossible to turn back, to accept that which was before, because to do so would be to die inside. Few, if any, forces in human affairs are as powerful as a shared vision of the heart.[1]

In its simplest intellectual form, a shared vision asks, What do we want to create? Beyond that, it becomes the focus and energy to actively create that which is desired. Few people, however, know what a vision, goals, or objectives are; how to create them; how to state them; or how to use them as guidelines for development.

A statement of vision is a general declaration that describes what a particular person, group of people, agency, or nation is striving to achieve. A vision is like a "vanishing point," the spot on the horizon where the straight, flat road on which you are driving disappears from view over a gentle rise in the distance. As long as you keep that vanishing point in focus as the place you want to go, you are free to take a few side trips down other roads and always know where you are in relation to where you want to go—your vision. It is therefore necessary to have at hand a dictionary and a thesaurus when crafting a vision statement because it must be as precise as possible. Through it, you must say what you mean and mean what you say.

Gifford Pinchot, the first chief of the US Forest Service, had a vision of protected forests that would produce commodities for people in perpetuity. In them, he saw the "greatest good for the greatest number in the long run."[2] Through his leadership, he inspired this vision, as a core value, around which everyone in the new agency could, and did, rally for almost a century.

I (Chris) spoke in 1989 to a First Nation of indigenous Canadians who owned a sawmill in central British Columbia. I had been asked to discuss how a coniferous forest functions, both above- and belowground, so that the First Nation could better understand the notion of productive sustainability, something of great concern to them. After I spoke, a contingent from the British Columbia provincial government told the indigenous Canadians what they could and could not do in the eyes of the government. The government officials were insensitive at best. The indigenous Canadians tried in vain to tell the officials how they *felt* about their land and how they were personally being treated. Both explanations fell on deaf ears.

After the meeting was over and the government people left, I explained to the indigenous Canadians what a vision is, why it is important, and how to create one. In this case, they already knew in their hearts what they wanted; they had a shared vision, but they could not articulate it in a way that the government people could understand because they dealt with the First Nation on a strictly intellectual basis.

Consequently, I helped them commit their feelings to paper as a vision statement for their sawmill in relation to the sustainable capacity of their land and their traditional ways. They were thus able to state their vision in a way that the government officials could understand, and it became their central point in future negotiations.

In another instance, I helped a president and vice president frame a vision, goals, and objectives for their new company. Although the president became frustrated during the 2-day process, he told me a couple of years later that it had been the most important exercise that he had ever been through for his company, and he uses it constantly as the company grows.

In contrast to a vision, a *goal* is a general statement of intent that remains until it is achieved, the need for it disappears, or the direction changes. Although a goal is a statement of direction that serves to further clarify the vision statement, it may be vague, and its accomplishment is not necessarily expected. A goal might be stated as, "My goal is to see Timbuktu."

There is, however, a saying in Nova Scotia for a person without a goal: "If you don't know where you're going, any path will take you there." Thus, without a goal, we take "potluck" in terms of where we will end up, which was Alice's dilemma when she met the Cheshire cat in Lewis Carroll's story *Alice's Adventures in Wonderland*. Alice asked the Cheshire cat:

> "Would you tell me, please, which way I ought to go from here?"
> "That depends a good deal on where you want to get to," said the Cat.
> "I don't much care where . . ." said Alice.
> "Then it doesn't matter which way you go," said the Cat.
> ". . . so long as I get somewhere," Alice added as an explanation.
> "Oh, you're sure to do that," said the Cat, "if you only walk long enough."[3]

An *objective*, on the other hand, is a specific statement of intended accomplishment. It is attainable, has a reference to time, is observable and measurable, and has an associated cost. The following are additional attributes of an objective: (1) It starts with an action verb; (2) it specifies a single outcome or result to be accomplished; (3) it specifies a date by which the accomplishment is to be completed; (4) it is framed in positive terms; (5) it is as specific and quantitative as possible and thus lends itself to evaluation; (6) it specifies only the "what, where," and "when" and avoids mentioning the "why" and the "how"; and (7) it is product oriented.

Let us consider the previous goal: My goal is to see Timbuktu. Now, let us make it into an objective: I *will* see Timbuktu on *my 21st birthday*. My stated objective is action oriented: I will see. It has a single outcome: seeing Timbuktu. It specifies a date, the day of my 21st birthday, and is framed in positive terms: I will. It lends itself to evaluation of whether or not I achieved my stated intent, and it clearly states *what, where*, and *when*, but *says nothing of the why or how*. Finally, it is product or outcome oriented: to see a specific place.

As you strive to achieve your objective, you must accept and remember that your objective is fixed, as though in concrete, but the plan to achieve your objective must remain flexible and changeable. A common human tendency, however, is to change the objective—devalue it—if it cannot be reached in the chosen way or by the chosen time. It is much easier, it seems, to devalue an objective than it is to change an elaborate plan that has shown it will not achieve the objective as originally conceived.

It is important to understand what is meant by a vision, goals, and objectives because they collectively tell us where we are going, the value of getting there, and the probability of success. Too often, however, we "sleeve shop." Sleeve shopping is going into a store to buy a jacket and deciding which jacket we like by the price tag on the sleeve.

The alternative to sleeve shopping is to first determine what you want by the perceived value and purpose of the outcome. Second, you must make the commitment to pay the price, whatever it is. Third, you must determine the price of achieving the outcome. Fourth, you must figure out how to fulfill your commitment—how to pay the price—and make a commitment to keep your commitment. Fifth, you must act on it.

Alexander the Great, the ancient Greek conqueror, provides an excellent example of knowing what one wants and how to achieve it. When he and his troops landed by ship on a foreign shore that he wanted to take, they found themselves badly outnumbered. As the story goes, he sent some men to burn the ships and then ordered his troops to watch the ships burn, after which he told them: "Now we win or die!"

Once you have completed your statement of vision, goals, and objectives, you will be able to answer the following questions concisely: (1) What do I want? (2) Why do I want it? (3) Where do I want it? (4) When do I want it? (5) From whom do I want it? (6) How much (or many) do I want? (7) For how long do I want it (or them)? If a single component is missing, you may achieve your desire by default, but not by design.

Only when you can concisely answer all of these questions are you ready for planning. Only then do you know where you want to go, know the value of going there, and can calculate the probability of arrival. Next, you must determine the cost, make the commitment to bear it and then commit yourself to keeping your commitment.

Although it is we who define our vision, goals, and objectives, it is the land that limits our options, and we must keep these limitations firmly in mind. At the same time, we must recognize that they can be viewed either as obstacles in our preferred path or as solid ground on which to build new paths. Remember, Nature deals in trends over various scales of time. Habitat (food, cover, water, space, and privacy) is a common denominator among species; we can use this knowledge to our benefit.[4] Long-term social-environmental sustainability requires that short-term economic goals and objectives be considered within the primacy of environmental postulates and sound, long-term ecological goals and objectives.[5]

How might such a process work? As an example, I (Chris) was asked to help the staff of a national forest in New Mexico come to grips with a vision, goals, and objectives for moving their forestry practices toward biological sustainability and hence economic sustainability. In so doing, I had to be very careful of my boundaries, which was sticky because I was, of necessity, wearing three hats.

I helped plan, participated in, and facilitated the outcome of a weeklong conference on the management of old-growth trees within a biologically sustainable forest. This conference was requested because of a long-standing and growing conflict, which was becoming increasingly destructive, both environmentally and culturally, among conservationists, three groups of Pueblo Indians, Latinos, Anglos, the timber industry, and the US Forest Service. The conflict was over old-growth forests, logging practices, traditional forest uses by both the indigenous Americans and the Latinos, and changes necessary for sustainable forestry.

The first 3 days were spent viewing slides that accompanied scientific presentations on how a forest and its streams function above- and below-ground (soils, water, trees, microbiology, mycology, animals, and people). Simultaneously with the ecological data, my task was to walk the audience through systems thinking and the consequences of decisions and how to ecologically and culturally link social-environmental sustainability.

The scientific presentations were followed by presentations of core values and points of view from each participating ethnic group. Throughout this period, there was free and open discussion among the collective group (over 50 people). During this time, people were asked to look at numbered color photographs on a wall and rank them from the most old-growth-like forest to the least. The participants were also asked to write a short description of what old growth was to them, including poetry if they wished.

The fourth day was spent in the field in an open discussion of several age classes of forest, wherein the participants verbally expressed which one felt the most "old growthy" to them and why. The point was for people to be able to express themselves in whatever way they were most comfortable and, in the process, convert the abstractions of the conference room into concrete social-environmental experiences.

The last day was again spent in the field, using a flip chart to come to grips with the notion of a collective vision and goals. Not only did a collective vision and goals gel but also the group agreed to rewrite the forest plan together. This was accomplished within a few months, as opposed to the year within which the forest supervisor had originally hoped to be able to revise the plan.

During this whole process, there were two detractors, both from the timber industry. Both of these gentlemen thought systems thinking (social-environmental sustainability) was "just philosophy" and beyond the purview of both science and such "emotional things" as visions. The rest of the group, however, outvoted them in the democratic process.

There was another gentleman, however, who purchased timber for one of the companies. After tentative beginnings, as the timber industry's representative on the planning committee, he became so caught up in the vision of biologically sustainable forestry that he departed from the industry's hardline stance. The power of the vision was so great for him that, when the company he worked for fired him because he changed his mind and heart, he said that while it was tough and frightening, he felt better about himself. His willingness to give up his job to accommodate his expanding consciousness is the kind of courage that continually strengthens my faith in humanity and in the power of transformative mediation.

What About the Children?

As we, the people, build our shared visions of a sustainable future in which each person's core values and expertise are acknowledged, we must exercise the good sense and humility to ask our children, beginning at least with second and third graders, what they think and how they feel about their future.

I (Chris) say this because I remember that, as a child, whenever I expressed my opinion about what I thought the world needed to be like or how I wanted it to be, I was inevitably invalidated as being "just a child who knew nothing" and was therefore summarily dismissed. But, consider for a moment that the children must inherit the world and its environment as we adults leave it for them. Our choices, our generosity or greed, our morality or licentiousness will determine the circumstances that will become their reality.

Why, then, do we adults assume that we know what is best for the children, their children, and their children's children when we are blatantly destroying their world with our blind greed and competitiveness? Why are children never asked what they expect of us as the caretakers and the trustees of the world they must inherit? Why are they never asked what they want us to leave them in terms of environmental quality? Why are they never asked what kinds of choices they would like to be able to make when they grow up? (For that matter, why do we not ask our elders where we have come from, how we are repeating history's mistakes, and what we have lost, such as fundamental human values discussed in Chapter 4?)

Where do we, the adults of the world, get the audacity to assume that we know what is good for our children when all over the world they are being abused at home by parents who are out of control. As well, children are being sold into slavery throughout the world for one reason or another and are being starved to death by adults using the allocation of food for political gain. We do not even know what is good for us. How can we possibly speak for them?

This lack of responsible care was keenly felt at the June 1992 worldwide Conference on the Environment held in Rio de Janeiro. A 12-year-old girl delivered to the entire delegation a most poignant speech about a child's perspective of the adult's environmental trusteeship. I (Chris) saw a video of the speech in which a child was pleading for a gentler hand on the environment so that there would be some things of value left for the children of the future. I saw an adult audience moved to tears—but not to action.

Times are changing. With the prevalence of technology, children are now galvanizing and actively utilizing adult mechanisms to move from the ranks of the disenfranchised. The youth have formed *Our Children's Trust* (https://www.ourchildrenstrust.org), a nonprofit organization. Exercising their constitutional rights by launching a legal campaign, they are bringing lawsuits against state agencies charged with protecting the environment. The plaintiffs urge the government to act swiftly to protect the ecosystem and the environment for them and future generations. They request immediate reduction of carbon and fossil fuel emissions and advocate for "legally-binding, science-based climate recovery policies."[6] To date, nine lawsuits are pending (in 9 states), while there are other proceedings in all 50 states and additional legal-youth action taking place in 13 other countries.[7]

In the society of the future, it is going to be increasingly important to listen to what the children say because they represent that which is to come. Children have beginner's minds. To them, all things are possible until adults with narrow minds, who have forgotten how to dream, put fences around their imaginations.

We adults, on the other hand, too often think we know what the answers should be and can no longer see what they might be. To us, whose imaginations were stifled by parents and schools, things have rigid limits of impossibility. We would do well, therefore, to consider carefully not only what the children say is possible in the future but also what they want. The future, after all, is theirs. (See Chapter 9, Ask the Children, in *The Perpetual Consequences of Fear and Violence: Rethinking the Future*[8] for a thorough discussion of what children want.)

Each generation must be the conscious keeper of the generation to come, not its judge. It is therefore incumbent on us, the adults, to prepare the way for those who must follow. This will entail, among other things, wise and prudent planning, beginning and ending with a firm commitment to the idea of social-environmental sustainability, in large measure through the resolution of ongoing environmental conflicts.

Discussion Questions

1. How is a destructive conflict transformed into a shared vision?
2. What does, "Who are we as a culture?" mean? Why is the question important?
3. Why is it important to have children involved in the mediation process? (How does the question relate to Biophysical Principles 1, 6–9 and 11?)
4. What are the differences among a vision, goals, and objectives?
5. Given our too-often short-term, current-generation-centered vision, how might we determine what is best for the children into the unforeseeable future?
6. Is there a particular question you would like to ask?

Endnotes

1. Robert D. Garrett. Mediation in Native America. *Dispute Resolution Journal*, March (1994):38–45.
2. Gifford Pinchot. *Breaking New Ground*. Harcourt, Brace, New York, 1947. 522 pp.
3. Lewis Carroll. *Alice's Adventures in Wonderland*. Merrill, New York, 1911. 187 pp.
4. Chris Maser. *Social-Environmental Planning: The Design Interface between Everyforest and Everycity*. CRC Press, Boca Raton, FL, 2009. 321 pp.
5. Cameron La Follette and Chris Maser. *Sustainability and the Rights of Nature: An Introduction*. CRC Press, Boca Raton, FL, 2017. 418 pp.
6. Our Children's Trust. Our Mission. https://www.ourchildrenstrust.org/mission-statement/ (accessed July 12, 2018).
7. Our Children's Trust. Other Proceedings in All 50 States. https://www.ourchildrenstrust.org/other-proceedings-in-all-50-states/ (accessed July 12, 2018).
8. Chris Maser. *The Perpetual Consequences of Fear and Violence: Rethinking the Future*. Maisonneuve Press, Washington, DC, 2004. 373 pp.

11

Modifying Our Belief Systems Regarding Change

This book opens with a poignant quest, "Do we humans living today owe anything to the current and future generations of life on Earth, both human and otherwise?" Through this inquiry, we have learned that our social-environmental future is governed by our past and current choices, much of which are shaped through our biophysical, emotional, intellectual, and spiritual perspectives. Fear-driven choices create unsustainable social-environment consequences. Because social-environmental aspects are invariably linked to Nature's biophysical principles, adhering to the latter can assist in realigning us with the natural functioning of Earth's systems.

The foregoing chapters offer guidance toward social-environmental cooperation for an environmental mediation practitioner and instructor. Provided are instruments and practices to help shift to more cooperative outcomes. Because changing behavior and attitudes can be frightening, mediators are provided with approaches to facilitate and equip disputing parties with tools to bring about their own transformation. This initiative, the choice to peaceful and lawful paths, builds community capacity to address future challenges for collective gain. In so doing, it requires the replacement of destructive behavior and attitudes with constructive choices and actions, signifying the death of harmful behavioral patterns.

In 1969, Elisabeth Kübler-Ross published a book, *On Death and Dying*,[1] which simultaneously is a book "on life and living." The author described five stages that a terminally ill person goes through when told of their impending death: denial and isolation, anger, bargaining, depression, and acceptance. Before relating these stages to our thought processes and how we change, let us examine each stage:

1. Denial, refusing to admit reality or trying to invalidate logic, is the first stage a terminally ill person goes through. Denial leads to a feeling of isolation, of being helpless and alone in the universe. At some level, however, the person knows the truth but is not yet ready or able to emotionally accept it.

2. Anger, which is a violent outward projection of fear, can be called *emotional panic*. The person is emotionally out of control because they now know they can no longer control circumstances.

3. Bargaining is when a person attempts to bargain with God to change the circumstances, to find a way out of having to deal with what is.

4. Depression is a somewhat different type of issue because it comes in two stages. In the first stage, a person is in the immediate process of losing control of circumstances, such as a job and their identity with that job. The second stage is one in which a person is no longer concerned with past losses, such as a job, but is taking impending losses into account, such as leaving loved ones behind.

5. Acceptance, the final stage, is creative and positive. With acceptance, returns trust, a faith in the goodness and justice of the outcome. Acceptance allows us to acknowledge our problem, which allows us to define it, which in turn allows us to transcend it.[2] But first, we must *accept what is*, which is *freedom* from fear.

Now, let us see how understanding these stages of dying help the living to understand dying *and* living. Although we are alive, we die daily to our ideas and belief systems, and in so doing, we go through the five stages of dying, which really are five stages of grieving. These stages are necessary as a process that prepares the way for change, a dying of the old thoughts and their attendant values and relationships within our lives to make way for the birth of the new:

1. Denial of or resistance to change is the first stage of a dying belief system in which we (in the generic) isolate ourselves because we see change as a condition to be avoided at almost any cost. We become defensive, fearful, and increasingly rigid in our thinking; we harden our attitudes and close our minds. If we become defensive, start to form a rebuttal before someone is finished speaking, and filter what is said to hear only what we want to hear, we are in denial.

2. Anger is the violent projection of uncontrollable fear. We are so afraid of change, of the dying of an old belief system, that we become temporarily insane: "I won't accept this!" Our anger, however, is not aimed at the person on whom it is projected; it is aimed at our own inability to control the circumstances that seem so threatening.

3. Bargaining is looking for a way to alter the circumstances based on more "acceptable" conditions, which is the purpose of such things as labor unions.

4. Depression is when we become resigned to our inability to control or change the "system," whatever that is, to suit our desires. We feel helpless and deliberately give up trying to alter circumstances. We become a "victim" of "outside forces," and our defense is to become cynical: distrustful of human nature and motives. A cynic is a critic who stresses faults and raises objections but assumes no

responsibility. A cynic sees the situation as hopeless and is therefore a prophet of doom who espouses self-fulfilling prophecies of failure regardless of the effort invested in success.

5. Acceptance of *what is* allows us to transcend the purely emotional state of the material mind and reach an integrated point of logic. In doing so, we can define the problem and, in turn, transcend it. Acceptance of the problem, however, must come before a resolution is possible.

Why do we (in the generic) fear unwanted, uncontrollable change so much? We resist such change because we are committed to protecting our existing belief system. Even if it is no longer valid, it represents the safety of past knowledge, in which there are no unwelcome surprises. We try to take our safe past and project it into an unknown future by skipping the present, which represents change and holds both uncertainty and accountability.

When confronted with change, we try to control the thoughts of others by accepting what, to us, are "approved" thoughts and rejecting "unapproved" thoughts. We see such control as a defense against unwanted change. But, as author George Bernard Shaw said, "My own education operated by a succession of eye-openers each involving the repudiation of some previously held belief."[3]

Change is the death of an accepted, "tried-and-true" belief system through which we have coped with life and which has become synonymous with our identity and therefore our security. When we get too "comfortable" with our belief systems, we might think of the turtle, for which only two choices in life exist: pull its head into its shell, where in safety it starves to death; or stick its neck out and risk finding something to eat and live.

Dysfunctional communities and their organizations with vested interests tend to hide within their self-serving ways by systematically distorting information. Such distortions do not depend on deliberate falsifications by individuals. Instead, people who are competent, hardworking, and honest can sustain systematic distortions by merely carrying out their organizational roles in an uncritical—and therefore personally safe—manner. Unchecked by outside influences or the undeniable realities of catastrophic failures, organizational systems can sustain self-serving distortions, even though the potential for catastrophic consequences is significant.

A technological culture, such as ours, faces two choices: It can wait until catastrophic failures expose systemic deficiencies, distortions, and self-deceptions (the turtle with its head tucked into its shell), or it can provide social checks and balances to correct for systemic distortions prior to catastrophic failures (the turtle with its head outside its shell, risking a view of the world). The second, more desirable, alternative, however, requires the active involvement of independent people who must ask "uncomfortable" questions and pursue "unfavorable" inquiries. Without such initiatives, checks and balances are undermined, and catastrophic possibilities are likely to increase as the scope and power of organizational technology expands.[4]

As we move forward in social-environmental sustainability, remember that success or failure is a crisis of the will and the imagination, not the possibilities. Remember also that to protect the best of what we have right now in the present for the present *and* the future, we must all continually change our thinking and our behavior to some extent. Society's saving grace is that we have a choice in almost everything but choosing. In that, we have no choice; we must choose. Therefore, whatever needs to be done can be done— when enough people want it to be done and decide to do it. The choice is ours, the adults of the world, but the consequences belong to the children of all generations into the unforeseeable future. So, do we humans living today owe anything to the current and future generations of life on Earth, both human and otherwise?

Discussion Questions

1. If change is inevitable, why are we afraid to accept it and then to respond positively to the best of our ability? Does our fear have anything to do with Biophysical Principles 7, 8, and 9 in Chapter 5?

2. What do you choose? Why?

3. If you could replace a destructive behavior or attitude with a constructive choice of action, what would it be?

4. Is there a particular question you would like to ask?

Endnotes

1. Elisabeth Kübler-Ross. *On Death and Dying*. Macmillan, New York, 1969.
2. Ibid.
3. LibQuotes. George Bernard Shaw. https://libquotes.com/george-bernard-shaw/quote/lbm7r3k (accessed June 28, 2017).
4. Chris Maser. *Global Imperative: Harmonizing Culture and Nature*. Stillpoint, Walpole, NH, 1992. 267 pp.

Appendix: Common and Scientific Names of Plants and Animals

ALGAE

Algae	Protista
Diatoms	Heterokontophyta

TREES AND SHRUBS

Apple	*Malus domestica*
Bigleaf maple	*Acer macrophyllum*
Bristlecone pine (Great Basin)	*Pinus longaeva*
Douglas-fir	*Pseudotsuga menziesii*
Giant sequoia	*Sequoiadendron giganteum*
Norway spruce	*Picea abies*
Oregon ash	*Fraxinus latifolia*
Oregon white oak	*Quercus garryana*
Pacific madrone	*Arbutus menziesii*
Red alder	*Alnus rubra*
Sitka spruce	*Picea sitchensis*
Western hemlock	*Tsuga heterophylla*
Western redcedar	*Thuja plicata*

INVERTEBRATES
VIRUS

Rinderpest	*Morbillivirus* spp.

SPIDERS AND ALLIES

Spiders	Arachnida

INSECTS

Gnat	Nematocera

VERTEBRATES
FISH

Atlantic salmon	*Salmo salar*
Cutthroat trout	*Oncorhynchus clarki*
Fish	Actinopterygii
Salmon	*Salmo* spp.
Steelhead trout	*Oncorhynchus mykiss*

AMPHIBIANS

Spadefoot toad	*Scaphiopus* spp.

BIRDS

Cliff swallows	*Petrochelidon pyrrhonota*
California condor	*Gymnogyps californianus*
Golden eagle	*Aquila chrysaetos*
Northern spotted owl	*Strix occidentalis*
Ostrich	*Struthio camelus*
Pileated woodpecker	*Dryocopus pileatus*
Raven	*Corvus corax*
Roadrunner	*Geococcyx* spp.
Steller's jay	*Cyanocitta stelleri*
Varied thrush	*Ixoreus naevius*
Wilson's Warbler	*Wilsonia pusilla*
Winter wren	*Troglodytes troglodytes*
Wrens	Troglodytidae

MAMMALS

African (aka Cape) buffalo	*Syncerus caffer*
Bears	Ursidae
Black-tailed deer	*Odocoileus hemionus columbianus*
Cow	*Bos taurus*
Fox	*Vulpes* spp.
Giraffe	*Giraffa camelopardalis*
Hyenas	Hyaenidae
Lion	*Panthera leo*
Mice	Rodentia
North American elk	*Cervus elaphus*
Rabbits	*Sylvilagus* spp.
Raccoon	*Procyon lotor*
Rhinoceros	Rhinocerotidae
River otter	*Lontra canadensis*
Shrews	Soricidae
Whales	Cetacea
Wildebeest	*Connochaetes* spp.
Wolves	*Canis lupus*
Zebra	*Equus* spp.

Index

T - #0508 - 101024 - C0 - 234/156/15 - PB - 9781032475561 - Gloss Lamination